SpringerBriefs in Mathematics

Series Editors

Krishnaswami Alladi
Ping Zhang
Yuri Tschinkel
Dierk Schleicher
Loring W. Tu
Nicola Bellomo
Otmar Scherzer

Mikhail Lifshits

Lectures on Gaussian Processes

 Springer

Prof. Mikhail Lifshits
Department of Mathematics and Mechanics
St. Petersburg State University
St. Petersburg
Russia

ISSN 2191-8198   e-ISSN 2191-8201
ISBN 978-3-642-24938-9   e-ISBN 978-3-642-24939-6
DOI 10.1007/978-3-642-24939-6
Springer Heidelberg Dordrecht London New York

Library of Congress Control: 2011943344

Printed on acid-free paper

Springer is part of Springer Science+Business Media (www.springer.com)

Preface

Gaussian processes can be viewed as a far-reaching infinite-dimensional extension of classical normal random variables. Their theory is one of the most advanced fields in the probability science and presents a powerful range of tools for probabilistic modelling in various academic and technical domains such as Statistics, Forecasting, Finance, Information Transmission, Machine Learning—to mention just a few.

The objective of these lectures is to present a quick and condensed treatment of the core theory that a reader must understand in order to make his own independent contributions. The primary intended readership are Ph.D/Masters students and researchers working in pure or applied mathematics. The knowledge of basics in measure theory, functional analysis, and, of course, probability, is required for successful reading.

The first chapters introduce essentials of the classical theory of Gaussian processes and measures. The core notions of Gaussian measure, reproducing kernel, integral representation, isoperimetric property, large deviation principle are explained and illustrated by numerous thoroughly chosen examples. This part mainly follows my book "Gaussian Random Functions" but the chosen exposition style is different. The brevity being a priority for teaching and learning purposes, certain technical details and proofs are omitted, rendering approach less formal, more appropriate to the lecture notes than to a textbook.

Obviously, new issues that emerged during last decade are also present in the exposition. Inequalities related to correlation conjecture and to other extremal problems, the entropy approaches to evaluation of small deviation probabilities, expansions of Gaussian vectors, relations to the theory of linear operators, and links to quantization problems for random processes fit into this category.

The short lecture notes by no means aim to provide a complete account of immense research field in pure and applied mathematics related to Gaussian processes. A few indications on further possible reading are given in "Invitation to Further Reading".

In university teaching, one can build a one-semester advanced course upon these lectures. Such courses were given by the author in Russia (St. Petersburg State University), in France (Université Lille I), in Germany (TU Darmstadt), in Finland (Helsinki University of Technology) and in USA (Georgia Institute of Technology) during last years. I am grateful to all mentioned host institutions for opportunity to teach my favorite subject in their rooms.

My sincere thanks go to Armin Straub for taking enthusiastic notes which served as an early draft of this text, and to Alexei Khartov for careful reading of the manuscript.

Abstract

Gaussian processes can be viewed as a far-reaching infinite-dimensional extension of classical normal random variables. Their theory presents a powerful range of tools for probabilistic modelling in various academic and technical domains such as Statistics, Forecasting, Finance, Information Transmission, Machine Learning—to mention just a few. The objective of these Briefs is to present a quick and condensed treatment of the core theory that a reader must understand in order to make his own independent contributions. The primary intended readership are Ph.D./Masters students and researchers working in pure or applied mathematics. The first chapters introduce essentials of the classical theory of Gaussian processes and measures with the core notions of reproducing kernel, integral representation, isoperimetric property, large deviation principle. The brevity being a priority for teaching and learning purposes, certain technical details and proofs are omitted. The later chapters touch important recent issues not sufficiently reflected in the literature, such as small deviations, expansions, and quantization of processes. In university teaching, one can build a one-semester advanced course upon these Briefs.

Keywords Gaussian processes · Gaussian measures · Isoperimetric inequalities · Large deviations · Reproducing Kernel Hilbert Space (RKHS) · Small deviations

Contents

Lectures on Gaussian Processes 1
1   Gaussian Vectors and Distributions 1
    1.1   Univariate Objects 1
    1.2   Multivariate Objects 2
    1.3   Gaussian Objects in "Arbitrary" Linear Spaces 5
2   Examples of Gaussian Vectors, Processes and Distributions 6
3   Gaussian White Noise and Integral Representations 13
    3.1   White Noise: Definition and Integration 13
    3.2   Integral Representations 15
4   Measurable Functionals and the Kernel 22
    4.1   Main Definitions 22
    4.2   Factorization Theorem 26
    4.3   Alternative Approaches to Kernel Definition 32
5   Cameron–Martin Theorem 34
6   Isoperimetric Inequality 38
    6.1   Euclidean Space 38
    6.2   Euclidean Sphere 39
    6.3   Poincaré Construction 40
    6.4   Euclidean Space with Gaussian Measure 41
    6.5   General Linear Space with Gaussian Measure 43
    6.6   Concentration Principle 43
7   Measure Concavity and Other Inequalities 46
    7.1   Measure Concavity 46
    7.2   Dilations 49
    7.3   Correlation Conjecture 50
    7.4   Bounds for Shifted Measures 53
8   Large Deviation Principle 54
    8.1   Cramér-Chernoff Theorem and General Large
           Deviation Principle 54

    8.2   Large Deviation Principle for Gaussian Vector   56
    8.3   Applications of Large Deviation Principle.   58
9   Functional Law of the Iterated Logarithm.   61
    9.1   Classical Law of the Iterated Logarithm   61
    9.2   Functional Law of the Iterated Logarithm   62
    9.3   Some Extensions   68
    9.4   Strong Invariance Principle   70
    9.5   FLIL for Random Walk   71
10 Metric Entropy and Sample Path Properties   72
    10.1   Basic Definitions   72
    10.2   Upper Bounds   73
    10.3   Lower Bounds   77
    10.4   *GB*-Sets and *GC*-Sets   80
11 Small Deviations   80
    11.1   Definitions and First Examples   80
    11.2   Markov Case   81
    11.3   Direct Entropy Method   83
    11.4   Dual Entropy Method   88
    11.5   Duality of Metric Entropy   92
    11.6   Hilbert Space   93
    11.7   Other Results   95
12 Expansions of Gaussian Vectors   96
    12.1   Problem Setting   96
    12.2   Series of Independent Random Vectors.   96
    12.3   Construction of a Vector with Given Distribution   99
    12.4   Expansion of a Given Vector. 100
    12.5   Examples: Expansions of Wiener Process 101
    12.6   Linear Operators, Associated Gaussian Vectors,
           and Their Expansions 105
13 Quantization of Gaussian Vectors 106
    13.1   Problem Setting 106
    13.2   Quantization and Small Deviations. 107
14 Invitation to Further Reading. 110
References. 110

Index 119

Lectures on Gaussian Processes

1 Gaussian Vectors and Distributions

Theory of random processes needs a kind of normal distribution. This is why Gaussian vectors and Gaussian distributions in infinite-dimensional spaces come into play. By simplicity, importance and wealth of results, theory of Gaussian processes occupies one of the leading places in modern Probability.

1.1 Univariate Objects

A real random variable X is *normally distributed* or *Gaussian* with expectation $a \in \mathbb{R}$ and variance $\sigma^2 > 0$, if its distribution density with respect to Lebesgue measure is

$$p(x) = \frac{1}{\sqrt{2\pi}\sigma} \exp\left\{\frac{-x^2}{2\sigma^2}\right\}.$$

We denote this distribution $N(a, \sigma^2)$ and write $X \sim N(a, \sigma^2)$. A normal distribution with zero variance $N(a, 0)$ is just the distribution concentrated at a point a.

If $X \sim N(a, \sigma^2)$, the characteristic function and Laplace transform of X are given by

$$\mathbb{E}e^{itX} = \exp\left\{iat - \frac{\sigma^2 t^2}{2}\right\},$$

$$\mathbb{E}e^{tX} = \exp\left\{at + \frac{\sigma^2 t^2}{2}\right\}.$$

By using the formula for characteristic function it is easy to check *stability* property: if the variables $X_1 \sim N(a_1, \sigma_1^2)$ and $X_2 \sim N(a_2, \sigma_2^2)$ are independent, then $X_1 + X_2 \sim N(a_1 + a_2, \sigma_1^2 + \sigma_2^2)$.

M. Lifshits, *Lectures on Gaussian Processes*, SpringerBriefs in Mathematics, DOI: 10.1007/978-3-642-24939-6_1, © Mikhail Lifshits 2012

The family of normal variables and distributions is also invariant with respect to linear transformations: if $X \sim N(a, \sigma^2)$, then

$$cX + d \sim N(d + ca, c^2\sigma^2).$$

Expectation and variance of a normal random variable coincide with parameters of its distribution:

$$\mathbb{E}X = a, \quad \mathbb{V}arX = \sigma^2.$$

Among normal distributions, the *standard* normal distribution $N(0, 1)$ plays a special role. Its distribution function is denoted by $\Phi(r)$. In other words,

$$\Phi(r) = \frac{1}{\sqrt{2\pi}} \int_{-\infty}^{r} \exp\left\{\frac{-x^2}{2}\right\} dx.$$

Let us notice a fast decay of the tails of normal distribution at infinity:

$$\Phi(-r) = 1 - \Phi(r) \sim \frac{1}{\sqrt{2\pi}r} \exp\left\{\frac{-r^2}{2}\right\}, \quad \text{as } r \to \infty.$$

A variable having any normal distribution $N(a, \sigma^2)$ can be obtained from the standard one by a linear transformation $X \mapsto Y = \sigma X + a$.

1.2 Multivariate Objects

A random vector $X = (X_j)_{j=1}^n \in \mathbb{R}^n$ is called *standard Gaussian*, if its components are independent and have a standard normal distribution. The distribution of X has a density

$$p(x) = \frac{1}{(2\pi)^{n/2}} \exp\left\{\frac{-(x, x)}{2}\right\}, \quad x \in \mathbb{R}^n.$$

There exist two equivalent definitions of a general Gaussian vector in $\mathbb{R}^n$.

Definition 1.1 A random vector $Y \in \mathbb{R}^n$ is called *Gaussian*, if it can be represented as $Y = a + LX$, where X is a standard Gaussian vector, $a \in \mathbb{R}^n$, and $L : \mathbb{R}^n \mapsto \mathbb{R}^n$ is a linear mapping.

Definition 1.2 A random vector $Y \in \mathbb{R}^n$ is called Gaussian, if the scalar product (v, Y) is a normal random variable for each $v \in \mathbb{R}^n$.

Definition 1.1 easily yields Definition 1.2. Indeed,

$$(a + LX, y) = (a, y) + (X, L^*y) = (a, y) + \sum_{j=1}^{n} (L^*y)_j X_j$$

has a normal distribution due to stability property.

In the multivariate setting, a definition of Gaussian distribution through a particular form of the density makes no much sense, because in many cases (when the operator L is degenerated, i.e. its image does not coincide with $\mathbb{R}^n$) the density just does not exist.

We will stick to Definition 1.2 that is more convenient for further generalizations: in most interesting spaces there is no standard Gaussian vector X required in Definition 1.1.

Similarly to the univariate notation $N(a, \sigma^2)$, the family of n-dimensional Gaussian distributions also admits a reasonable parametrization. Recall that for any random vector $Z = (Z_j) \in \mathbb{R}^n$ one understands the expectation component-wise, i.e. $\mathbb{E}Z = (\mathbb{E}Z_j)$, while its *covariance operator* $K_Z : \mathbb{R}^n \mapsto \mathbb{R}^n$ is defined by

$$cov((v_1, Z), (v_2, Z)) = (v_1, K_Z v_2).$$

If all components of a vector Z have finite second moments, then the expectation $\mathbb{E}Z$ and covariance operator K_Z do exist. There is no restrictions on expectation value, while the covariance operator is necessarily non-negative definite and symmetric. In other words, there exists an orthonormal base (e_j) such that K has a diagonal form $K_Z e_j = \lambda_j e_j$ with $\lambda_j \geq 0$.

We write $Y \sim N(a, K)$ if Y is a Gaussian vector with expectation a and covariance operator K. In particular, for a standard Gaussian vector we have $X \sim N(0, E_n)$, where $E_n : \mathbb{R}^n \mapsto \mathbb{R}^n$ is the identity operator.

The suggested notation generates legitimate questions:

- Does $N(a, K)$ exist for all $a \in R^n$ and all non-negative definite and symmetric operators K ?
- Is the distribution $N(a, K)$ unique ?
- Is it true that any Gaussian distribution has a form $N(a, K)$?

Let us answer positively on all these questions. Indeed, let $a \in R^n$, and let K be a non-negative definite and symmetric linear operator. Consider a base (e_j) corresponding to the diagonal form of K (see above) and define $L = K^{1/2}$ by relations $Le_j = \lambda_j^{1/2} e_j$. Consider a random vector $Y = a + LX$. We have already seen that Y is Gaussian. It is almost obvious that it has expectation a and covariance operator K.

The uniqueness of $N(a, K)$ follows from the fact that a pair (a, K) determines the distribution of (v, Y) as $N((v, a), (v, Kv))$, hence by classical Cramér–Wold theorem the entire distribution is determined uniquely. Notice by the way that the distribution of (v, Y) yields a formula for the characteristic function

$$\mathbb{E}e^{i(v, Y)} = \exp\left\{ i(v, a) - \frac{(v, Kv)}{2} \right\}.$$

Finally, all components of a Gaussian vector are normal random variables, hence they have finite second moments. Therefore, any Gaussian vector has an expectation and a covariance operator, i.e. any Gaussian distribution can be written in the form $N(a, K)$.

Exercise 1.1 Assume that a vector Y satisfies Definition 1.2. Prove that it also satisfies Definition 1.1.

Exercise 1.2 Let all components of a random vector Y be normal random variables. Does it follow that Y is a Gaussian vector?

As in the one-dimensional case, the Gaussian property is preserved by summation of independent random vectors (stability property) and by a linear transformation. If the vectors $X_1 \sim N(a_1, K_1)$ and $X_2 \sim N(a_2, K_2)$ are independent, then $X_1 + X_2 \sim N(a_1 + a_2, K_1 + K_2)$.

If $L : \mathbb{R}^n \mapsto \mathbb{R}^n$ is a linear operator, $h \in \mathbb{R}^n$, and $X \sim N(a, K)$, then

$$LX + h \sim N(h + La, LKL^*).$$

Norm distribution of a Gaussian vector

Let $X = (X_j)_{j=1}^n \in \mathbb{R}^n$ be a standard Gaussian vector. The density formula yields

$$\mathbb{P}\{\|X\| \le R\} = c \int_0^R r^{n-1} \exp\{-r^2/2\}dr,$$

with a normalizing factor $c = 2^{1-n/2} \Gamma(n/2)^{-1}$. We also know that

$$\mathbb{E}\|X\|^2 = \sum_{j=1}^n \mathbb{E}X_j^2 = n.$$

Moreover, if n is *large*, a substantial part of mass of the standard Gaussian distribution is concentrated in a band of constant width around $\sqrt{n}$. Indeed, we can apply the law of large numbers and the central limit theorem to the sum

$$\|X\|^2 = \sum_{j=1}^n X_j^2,$$

thus

$$\frac{1}{n} \sum_{j=1}^n X_j^2 \Rightarrow 1, \quad \text{hence} \quad \frac{\|X\|}{\sqrt{n}} \Rightarrow 1.$$

For a band of width r we have

$$\mathbb{P}\{|\|X\| - \sqrt{n}| \le r\}$$

$$= \mathbb{P}\left\{\frac{(\sqrt{n}-r)^2 - n}{\sqrt{n}} \le \frac{\sum_{j=1}^n X_j^2 - n}{\sqrt{n}} \le \frac{(\sqrt{n}+r)^2 - n}{\sqrt{n}}\right\}$$

$$\to 2\Phi\left(\frac{2r}{\sqrt{\mathbb{V}ar(X_j^2)}}\right) - 1 = 2\Phi\left(\sqrt{2}r\right) - 1.$$

These calculations show that *in high-dimensional spaces the standard Gaussian distribution is similar to the uniform distribution on the sphere of a corresponding radius.*

1.3 Gaussian Objects in "Arbitrary" Linear Spaces

Let $\mathscr{X}$ be a linear topological space (its additional required properties will be mentioned below). Let $\mathscr{X}^*$ denote the dual space of continuous linear functionals on $\mathscr{X}$. We denote (f, x) the duality between the spaces $\mathscr{X}$ and $\mathscr{X}^*$, i.e. (f, x) stands for the value of a functional $f \in \mathscr{X}^*$ on an element $x \in \mathscr{X}$. A random vector X taking values in $\mathscr{X}$ is a measurable mapping $X : (\Omega, \mathscr{F}, \mathbb{P}) \mapsto \mathscr{X}$. A σ-field on $\mathscr{X}$ should be sufficiently large to provide measurability of all continuous linear functionals.

Gaussian vectors, their expectations and covariance operators are defined in a same way as in the finite-dimensional case.

A random vector $X \in \mathscr{X}$ is called *Gaussian*, if (f, x) is a normal random variable for all $f \in \mathscr{X}^*$.

A vector $a \in \mathscr{X}$ is called *expectation* of a random vector $X \in \mathscr{X}$, if $\mathbb{E}(f, X) = (f, a)$ for all $f \in \mathscr{X}^*$. We write $a = \mathbb{E}X$ in this case. A linear operator $K : \mathscr{X}^* \mapsto \mathscr{X}$ is called *covariance operator* of a random vector $X \in \mathscr{X}$, if $cov((f_1, X), (f_2, X)) = (f_1, Kf_2)$ for all $f_1, f_2 \in \mathscr{X}^*$. We write $K = cov(X)$ in this case. Covariance operator is symmetric,

$$(f, Kg) = (g, Kf), \quad \forall f, g \in \mathscr{X}^*,$$

and non-negative definite, i.e.

$$(f, Kf) \geq 0, \quad \forall f \in \mathscr{X}^*.$$

From the definition of Gaussian vector, we see that it only makes sense when the space of continuous linear functionals on $\mathscr{X}$ is rich enough. For example, if $\mathscr{X}^* = \{0\}$, then any vector satisfies this definition rendering it senseless. Therefore, usually one of three situations of increasing generality is considered.

(1) $\mathscr{X}$ is a separable Banach space, for example, $\mathbb{C}[0, 1]$, $L_p[0, 1]$ etc.
(2) $\mathscr{X}$ is a complete separable locally convex metrizable topological linear space, for example, $\mathbb{C}[0, \infty)$, $\mathbb{R}^\infty$ etc.
(3) $\mathscr{X}$ is a locally convex linear topological space and a vector X is such that its distribution is a Radon measure.

In cases (1) and (2) every finite measure is a Radon measure, thus case (3) is the most general one.

In the subsequent exposition, we always assume by default that one of these assumptions is satisfied (i.e. at least assumption (3) holds), and call them *usual assumptions.*

As in finite-dimensional case, we assert that X has a distribution $N(a, K)$, if X is a Gaussian vector with expectation a and covariance operator K.

The same questions arise again:

- Does $N(a, K)$ exist for all $a \in \mathscr{X}$ and all symmetric non-negative definite operators $K : \mathscr{X}^* \mapsto X$?
- Is the distribution $N(a, K)$ unique?
- Is it true that any Gaussian distribution has a form $N(a, K)$?

The answers will be slightly different from those given in the previous subsection. As for the first question, the existence of $N(a, K)$ depends only on K. Indeed, if a random vector X has a distribution $N(a_1, K)$, then the vector $X + a_2 - a_1$ has a distribution $N(a_2, K)$. On the other hand, the following exercise shows that the distribution $N(0, K)$ does not necessarily exist for a symmetric non-negative definite operator K.

Exercise 1.3 Let $\mathscr{X}$ be an infinite-dimensional separable Hilbert space. Then $\mathscr{X}^* = \mathscr{X}$ and identity operator $E : \mathscr{X} \mapsto \mathscr{X}$ is a symmetric non-negative definite operator. Prove that the distribution $N(0, E)$ does not exist. Hint: would a random vector X have distribution $N(0, E)$, it would satisfy an absurd identity $\mathbb{P}(||X||^2 = \infty) = 1$.

Finding a criterion for existence of $N(0, K)$ is highly non-trivial problem, and the solution depends on the space $\mathscr{X}$. This question is deliberately omitted in these lectures (except for the Hilbert space case), because we are rather interested in investigation of objects that certainly exist.

Fortunately, *under usual assumptions* we can give positive answers on two remaining questions. Namely, every Gaussian vector possesses an expectation and a covariance operator, see [117] for details. Therefore, its distribution belongs to the family $\{N(a, K)\}$.

Furthermore, a pair (a, K) determines the distribution of a variable (f, x) as $N((f, a), (f, Kf))$, and we find the characteristic function

$$\mathbb{E}e^{i(f, X)} = \exp\left\{i(f, a) - \frac{(f, Kf)}{2}\right\}.$$

Any Radon distribution in $\mathscr{X}$ is determined by its characteristic function. Therefore, distribution $N(a, K)$ is unique.

2 Examples of Gaussian Vectors, Processes and Distributions

Example 2.1 (Standard Gaussian measure in $\mathbb{R}^\infty$) Consider the space of all sequences $\mathbb{R}^\infty$ equipped with topology of coordinate convergence. It becomes a complete separable metric space by introducing an appropriate distance, e.g.

$$\rho(x, y) = \sum_{j=1}^{\infty} 2^{-j} \min\{|x_j - y_j|, 1\}.$$

Thus, our "usual assumptions" are satisfied. Recall that the dual space $\mathscr{X}^* = \mathbf{c_0}$ is a space of all *finite* sequences, and the duality is

$$(f, x) = \sum_{j=1}^{\infty} f_j x_j,$$

where the sum is in fact a finite one. Consider a sequence of i.i.d. $N(0, 1)$-distributed random variables as a vector $X \in \mathscr{X}$. Due to stability of normal distribution, for any $f \in \mathscr{X}^*$ the random variable (f,X) is $N(0, \sigma^2)$-distributed with variance $\sigma^2 = \sum_{j=1}^{\infty} f_j^2$. Therefore, X is a Gaussian vector. It is clear that $\mathbb{E}X = 0$. Embedding operator $K : \mathbf{c_0} \mapsto \mathbb{R}^{\infty}$ serves as covariance operator for X. Indeed,

$$cov((f, X)(g, X)) = \mathbb{E}(f, X)(g, X) = \mathbb{E}\left(\sum_{j=1}^{\infty} f_j X_j\right)\left(\sum_{j=1}^{\infty} g_j X_j\right)$$

$$= \sum_{j_1=1}^{\infty} \sum_{j_2=1}^{\infty} f_{j_1} g_{j_2} \mathbb{E}\left(X_{j_1} X_{j_2}\right) = \sum_{j=1}^{\infty} f_j g_j = (f, Kg).$$

We call the distribution of X a *standard Gaussian measure* in $\mathbb{R}^{\infty}$.

Example 2.2 (Gaussian vectors in a Hilbert space) Let $\mathscr{X}$ be a separable Hilbert space whose scalar product will be denoted by $(\cdot, \cdot)$. Then we may identify $\mathscr{X}^*$ with $\mathscr{X}$. The "usual assumptions" are clearly satisfied. For building a Gaussian vector in $\mathscr{X}$ we need: an orthonormal base (e_j) in $\mathscr{X}$, a sequence of independent $N(0, 1)$-distributed random variables (ξ_j), and a sequence of non-negative numbers (σ_j) satisfying assumption $\sum_{j=1}^{\infty} \sigma_j^2 < \infty$. Then we define X by the formula

$$X = \sum_{j=1}^{\infty} \sigma_j \xi_j e_j, \tag{2.1}$$

where the series is a.s. convergent in Hilbert norm of $\mathscr{X}$. This representation is called *Karhunen–Loève expansion.*

For any $f = \sum_j f_j e_j \in \mathscr{X}$ the random variable

$$(f, X) = \sum_{j=1}^{\infty} \sigma_j f_j \xi_j$$

is normally distributed with zero mean and variance $\sum_{j=1}^{\infty} \sigma_j^2 f_j^2$. Therefore, X is a Gaussian random vector and $\mathbb{E}X = 0$. In order to find its covariance operator, let us compute

$$cov((f, X)(g, X)) = \mathbb{E}(f, X)(g, X) = \mathbb{E}\left(\sum_{j=1}^{\infty} \sigma_j f_j \xi_j\right)\left(\sum_{j=1}^{\infty} \sigma_j g_j \xi_j\right)$$

$$= \sum_{j_1=1}^{\infty} \sum_{j_2=1}^{\infty} f_{j_1} \sigma_{j_1} g_{j_2} \sigma_{j_2} \mathbb{E}\left(\xi_{j_1} \xi_{j_2}\right) = \sum_{j=1}^{\infty} \sigma_j^2 f_j g_j.$$

Therefore,

$$(f, Kg) = \sum_{j=1}^{\infty} \sigma_j^2 f_j g_j.$$

By plugging in the base elements, we find that

$$K : g \mapsto \sum_{j=1}^{\infty} \sigma_j^2 g_j e_j.$$

In other words, K is a diagonal operator with respect to the base e_j and σ_j^2 are the corresponding eigenvalues.

One can show that *any* Gaussian vector in a Hilbert space admits a representation (2.1), see [157]. This means that a Gaussian distribution with covariance operator K exists iff, in appropriate base, K has a diagonal form with non-negative eigenvalues, and the sum of these eigenvalues is finite.

Exercise 2.1 Prove the convergence of series (2.1) that defines vector X.

Further examples of Gaussian vectors and distributions are related to the notion of Gaussian random process. Recall that a *random process* X on a parametric set T is a family of random variables $X(t, \omega), t \in T$, defined on a common probability space $(\Omega, \mathscr{F}, \mathbb{P})$. A process X is called *Gaussian* if for any $t_1, \ldots, t_n \in T$ the distribution of the random vector $(X(t_1), \ldots, X(t_n))$ is a Gaussian distribution in $\mathbb{R}^n$. The properties of a Gaussian process are completely determined by its expectation $\mathbb{E}X(t), t \in T$, and covariance $cov(X(s), X(t)), s, t \in T$.

If T is a topological space, we say that X has continuous sample paths, if the function $X(\cdot, \omega)$ is continuous on T for $\mathbb{P}$-almost every $\omega \in \Omega$.

Example 2.3 (*Wiener process*) Let $\mathscr{X} = \mathbb{C}[0, 1]$ be the Banach space of all continuous functions on the interval $[0, 1]$ equipped with the supremum norm

$$||x|| = \max_{t \in [0,1]} |x(t)|$$

and with the corresponding topology of uniform convergence. The dual space $\mathscr{X}^* = \mathbb{M}[0, 1]$ is the space of charges (sign measures) of finite variation on $[0, 1]$. The duality is given by

$$(\mu, f) = \int_{[0,1]} f d\mu, \quad \mu \in \mathbb{M}[0, 1], \ f \in \mathbb{C}[0, 1].$$

We will now consider a Gaussian vector composed of the sample paths of a *Wiener process* $X = W(t)$, $0 \leq t \leq 1$, i.e. of a process satisfying assumptions

$$\mathbb{E}W(t) = 0, \qquad \mathbb{E}W(s)W(t) = \min\{s, t\}.$$

Let us find the expectation and covariance operator of W. Since

$$\mathbb{E}(\mu, W) = \mathbb{E}\int_{[0,1]} W d\mu = \int_{[0,1]} \mathbb{E}W(t)\mu(dt) = 0,$$

we have $\mathbb{E}W = 0$. Moreover,

$$cov((\mu, W), (v, W)) = \mathbb{E}(\mu, W)(v, W) = \mathbb{E}\int_{[0,1]} W d\mu \int_{[0,1]} W dv$$

$$= \mathbb{E}\int_{[0,1]^2} W(s)W(t)\mu(ds)v(dt)$$

$$= \int_{[0,1]^2} \mathbb{E}W(s)W(t)\mu(ds)v(dt)$$

$$= \int_{[0,1]^2} \min\{s, t\}\mu(ds)v(dt).$$

Therefore,

$$(\mu, Kv) = \int_{[0,1]^2} \min\{s, t\}\mu(ds)v(dt),$$

and we find that

$$(Kv)(s) = \int_{[0,1]} \min\{s, t\}v(dt).$$

Recall the basic properties of a Wiener process.

- It is $1/2$-*self-similar*, i.e. for any $c > 0$ the process $Y(t) := \frac{W(ct)}{\sqrt{c}}$ is also a Wiener process;
- It has stationary increments;
- It has independent increments;
- It is a Markov process;
- It admits time inversion: the process $Z(t) := tW(\frac{1}{t})$ is also a Wiener process.

To a large extent, the importance of Wiener process is explained by its fundamental role in stochastic calculus and in the limit theorems for random processes (invariance principle).

A careful reader will notice that we did not use any special properties of Wiener process (except for sample path continuity $W \in \mathbb{C}[0, 1]$), while computing its covariance. Therefore, we can extend the previous example to the following one.

Example 2.4 (*Arbitrary continuous Gaussian process*) Let T be a compact metric space, let $\mathscr{X} = \mathbb{C}(T)$ denote the Banach space of all continuous functions on T equipped with supremum norm $||x|| = \max_{t \in T} |x(t)|$ and with the corresponding topology of uniform convergence. The dual space $\mathscr{X}^* = \mathbb{M}(T)$ is a space of charges (sign measures) on T. The duality is given by

$$(\mu, f) = \int_T f d\mu, \quad \mu \in \mathbb{M}(T), \ f \in \mathbb{C}(T).$$

Let $X(t), t \in T$, be a Gaussian random process with continuous sample paths on the parametric set T. It is completely characterized by the functions

$$a(t) := \mathbb{E}X(t), \qquad K(s, t) := cov(X(s), X(t)).$$

Then we can view at X as a random element of $\mathscr{X}$, while its expectation and covariance operator K can be calculated by $\mathbb{E}X = a$ and

$$(K\nu)(s) = \int_T K(s, t)\nu(dt). \tag{2.2}$$

We present now several most interesting Gaussian processes fitting into this scheme.

Example 2.5 (*Fractional Brownian motion*) Let $\alpha \in (0, 2]$. A Gaussian process $W^{(\alpha)}(t), t \in \mathbb{R}$, is called an α-*fractional Brownian motion* (*fBm*), if

$$\mathbb{E}W^{(\alpha)}(t) = 0, \quad \mathbb{E}W^{(\alpha)}(s)W^{(\alpha)}(t) = \frac{1}{2}\left(|s|^\alpha + |t|^\alpha - |s - t|^\alpha\right).$$

The choice of covariance function might seem strange but there is a more natural equivalent definition

$$\mathbb{E}W^{(\alpha)}(t) = 0, \quad W^{(\alpha)}(0) = 0, \quad \mathbb{E}\left|W^{(\alpha)}(s) - W^{(\alpha)}(t)\right|^2 = |s - t|^\alpha. \tag{2.3}$$

Notice that for $\alpha = 1$ we have a classical Wiener process. If $\alpha = 2$, we obtain a quite degenerated process with linear sample paths entirely determined by their values at a single moment: $W^{(2)}(t) = t W^{(2)}(1)$.

Recall the basic properties of fractional Brownian motion which can be easily derived from (2.3).

- It is H-self-similar for $H = \frac{\alpha}{2}$. This means that for any $c > 0$ the process $Y(t) := \frac{W^{(\alpha)}(ct)}{c^H}$ is also an α-fractional Brownian motion;
- It has stationary increments;
- Its increments are dependent (except for the Wiener case $\alpha = 1$);
- It is not a Markov process (except for the Wiener case $\alpha = 1$);
- For $\alpha \in (1, 2]$ it has a long range dependence property which will not be discussed here.

Fractional Brownian motion plays an important role in limit theorems for random processes as well as the Wiener process, especially in the long range dependence case.

Exercise 2.2 Check the equivalence of two definitions of fBm.

Exercise 2.3 Find a limit for covariance function of α-fractional Brownian motion, as $\alpha \to 0$. How should one construct a process with the limiting covariance function?

Example 2.6 (*Gaussian Markov processes*) It is known that covariance function of a Gaussian Markov process has a form

$$K(s, t) = A(\min\{s, t\})\, B(\max\{s, t\}). \qquad (2.4)$$

The processes with covariances of this type can be easily derived from a Wiener process. Let $X(t) = f(t)W(g(t))$, where $g(\cdot)$ is an increasing function. Then

$$\mathbb{E}X(s)X(t) = f(s)f(t)\min\{g(s), g(t)\}$$

has a form (2.4) with $A = fg$, $B = f$. Conversely, for given A, B let $f = B$, $g = A/B$. Here are some examples of (2.4): $\min\{s, t\}$ (corresponds to a Wiener process),

$$\min\{s, t\} - st = \min\{s, t\}(1 - \max\{s, t\})$$

(corresponds to *Brownian bridge*) and

$$e^{-|s-t|/2} = e^{\min\{s,t\}/2}e^{-\max\{s,t\}/2}$$

(corresponds to *Ornstein–Uhlenbeck process*). The latter process is an example of Gaussian stationary process and has a very short memory in the sense that it can be written in the form

$$X(t) = e^{-t/2}X(0) + V(t), \quad t \geq 0,$$

with $V(t)$ independent of the past $\{X(s), s \leq 0\}$. The spectral representation of its covariance involves Cauchy measure, i.e.

$$e^{-|t|/2} = \frac{2}{\pi} \int_{-\infty}^{\infty} e^{itv} \frac{dv}{1 + 4v^2}.$$

Example 2.7 (*Brownian sheet or Wiener–Chentsov random field*) A Gaussian process $W(t), t \in \mathbb{R}_+^d$, is called *Brownian sheet* or *Wiener–Chentsov field*, if

$$\mathbb{E}W(t) = 0, \quad \mathbb{E}W(s)W(t) = \prod_{l=1}^{d} \min\{s_l, t_l\}. \qquad (2.5)$$

For $d = 1$ we obtain a classical Wiener process.

Correlation function of Brownian sheet admits a beautiful geometric interpretation. Let us connect with any point $t \in \mathbb{R}_+^d$ a parallelepiped

$$[0, t] := \{s \in \mathbb{R}^d : 0 \le s_l \le t_l, 1 \le l \le d\}.$$

Then

$$\prod_{l=1}^{d} \min\{s_l, t_l\} = \lambda^d \left([0, s] \cap [0, t]\right),$$

where λ^d denotes Lebesgue measure in $\mathbb{R}_+^d$.

It easy to deduce from (2.5) that $W(t)$ is H-self-similar for $H = \frac{d}{2}$. This means that for any $c > 0$ the process $Y(t) := \frac{W(ct)}{c^H}$ is also a Brownian sheet.

Similarly to Wiener process, Brownian sheet possesses certain property of "independent increments" that we will not state here.

Brownian sheet is a special case of *tensor product* of random processes, which is a random field with covariance

$$K(s, t) = \prod_{l=1}^{d} K_l(s_l, t_l),$$

where $K_l(\cdot, \cdot)$ are covariance functions of one-parametric processes that do not necessarily coincide with each others. For example, a famous *Kiefer field* is a tensor product of Wiener process and Brownian bridge, i.e its covariance is equal to

$$K(s, t) = \min\{s_1, t_1\} \cdot (\min\{s_2, t_2\} - s_2 t_2), \quad s_1, t_1 \ge 0, \ 0 \le s_2, t_2 \le 1.$$

Kiefer field emerges as a limit of normalized empirical distribution function taking time into account. Recall that this function is defined for a sample $\{X_i, 1 \le i \le n\}$ of independent random variables uniformly distributed on $[0, 1]$ by

$$F_n(t, r) = \sqrt{n} \left(\frac{\#\{i : X_i \le r, i \le tn\}}{n} - rt \right).$$

As $n \to \infty$, the random fields $F_n(\cdot, \cdot)$ converge to Kiefer field in distribution.

Example 2.8 (*Lévy's Brownian function* [110]) A Gaussian process $W^L(t)$, $t \in \mathbb{R}^d$, is called *Lévy's Brownian function* or *Lévy field* if

$$\mathbb{E} W^L(t) = 0, \quad \mathbb{E} W^L(s) W^L(t) = \frac{1}{2} \left(\|s\| + \|t\| - \|s - t\| \right). \tag{2.6}$$

Here $\|\cdot\|$ stands for the usual Euclidean norm in $\mathbb{R}^d$.

For $d=1$ Lévy's Brownian function corresponds to a pair of independent Wiener processes (for $t \geq 0$ and for $t \leq 0$).

Similarly to the one-parametric case, there is an equivalent definition of the form

$$\mathbb{E}W^L(t) = 0, \quad W^L(0) = 0, \quad \mathbb{E}\left|W^L(s) - W^L(t)\right|^2 = ||s - t||. \qquad (2.7)$$

It is easy to observe from (2.6) that $W^L(t)$ is an H-self-similar process for $H = \frac{1}{2}$. This means that for any $c > 0$ the process $Y(t) := \frac{W^L(ct)}{c^{1/2}}$ is also a Lévy's Brownian function.

Exercise 2.4 Examples 2.7 and 2.8 extend the notion of Wiener process to the case of d-parametric random fields. State similar extensions for α-fractional Brownian motion with arbitrary $\alpha \in (0, 2)$. Explore the self-similarity properties of extended processes.

Example 2.9 (Bifractional Brownian motion [83]) A Gaussian process $W^{\alpha,K}(t), t \in \mathbb{R}_+$, is called (α, K)-*bifractional Brownian motion* if

$$\mathbb{E}W^{\alpha,K}(t) = 0, \quad \mathbb{E}W^{\alpha,K}(s)W^{\alpha,K}(t) = \frac{1}{2^K}\left((t^\alpha + s^\alpha)^K - |t-s|^{\alpha K}\right), \quad t, s \geq 0.$$

Note that letting $K = 1$ yields a usual fractional Brownian motion W^α. The process $W^{\alpha,K}$ exists provided that $0 < \alpha \leq 2$, $0 < K \leq 2$, and $\alpha K \leq 2$. See Exercise 3.2 in the next section for the relation between bifractional and ordinary fractional Brownian motion.

3 Gaussian White Noise and Integral Representations

3.1 White Noise: Definition and Integration

Many Gaussian random functions admit a convenient definition or representation by means of white noise integral.

Let $(\mathcal{R}, \mathcal{A}, \nu)$ be a measure space. Let $\mathcal{A}_0 = \{A \in \mathcal{A} : \nu(A) < \infty\}$. Gaussian random function $\{\mathcal{W}(A), A \in \mathcal{A}_0\}$ is called *Gaussian white noise* with control measure ν if $\mathbb{E}\mathcal{W}(A) = 0$ and $\mathbb{E}\mathcal{W}(A)\mathcal{W}(B) = \nu(A \cap B)$. The main properties of Gaussian white noise are as follows:

- $\mathbb{V}ar\,\mathcal{W}(A) = \nu(A)$;
- If the sets $A_1, \ldots, A_n$ are disjoint, then the variables $\mathcal{W}(A_1), \ldots, \mathcal{W}(A_n)$ are independent;
- If the sets $A_1, \ldots, A_n$ are disjoint, then

$$\sum_{j=1}^n \mathcal{W}(A_j) = \mathcal{W}\left(\bigcup_{j=1}^n A_j\right) \quad \text{a.s.}$$

Exercise 3.1 Deduce these properties from the definition.

For any function $f \in L_2(\mathscr{R}, \mathscr{A}, \nu)$, we will define a *white noise integral* $\int_{\mathscr{R}} f \, \mathrm{d}\mathscr{W}$. First, we do it for step functions by

$$\int_{\mathscr{R}} \left(\sum_j c_j \mathbf{1}_{A_j} \right) \mathrm{d}\mathscr{W} := \sum_j c_j \mathscr{W}(A_j), \quad c_j \in \mathbb{R}, A_j \in \mathscr{A}_0,$$

and check the correctness of this definition, i.e.

$$\sum_j c_j \mathbf{1}_{A_j} = \sum_i b_i \mathbf{1}_{B_i} \curvearrowright \sum_j c_j \mathscr{W}(A_j) = \sum_i b_i \mathscr{W}(B_i).$$

Next, in the class of step functions we establish linearity

$$\int_{\mathscr{R}} (cf) \mathrm{d}\mathscr{W} = c \int_{\mathscr{R}} f \, \mathrm{d}\mathscr{W}; \quad \int_{\mathscr{R}} (f + g) \mathrm{d}\mathscr{W} = \int_{\mathscr{R}} f \, \mathrm{d}\mathscr{W} + \int_{\mathscr{R}} g \, \mathrm{d}\mathscr{W}$$

and isometric property

$$\mathbb{E} \left(\int_{\mathscr{R}} f \, \mathrm{d}\mathscr{W} \cdot \int_{\mathscr{R}} g \, \mathrm{d}\mathscr{W} \right) = \int_{\mathscr{R}} fg \, \mathrm{d}\nu,$$

which yields, in particular,

$$\mathbb{V}ar \int_{\mathscr{R}} f \, \mathrm{d}\mathscr{W} = \int_{\mathscr{R}} |f|^2 \mathrm{d}\nu = \|f\|_2^2.$$

Clearly,

$$\mathbb{E} \int_{\mathscr{R}} f \, \mathrm{d}\mathscr{W} = 0.$$

Since the class of step functions is dense in $L_2(\mathscr{R}, \mathscr{A}, \nu)$, for any function $f \in L_2(\mathscr{R}, \mathscr{A}, \nu)$ we can define its integral as an L_2-limit,

$$\int_{\mathscr{R}} f \, \mathrm{d}\mathscr{W} := \lim_{n \to \infty} \int_{\mathscr{R}} f_n \mathrm{d}\mathscr{W}.$$

where (f_n) is any sequence of step functions converging to f in L_2. Due to isometric property the limit exists and does not depend on the sequence (f_n). There is no problem to transfer the above mentioned properties of the integral from the class of step functions to the entire space L_2.

Complex white noise

Complex-valued Gaussian white noise and the respective integral are defined along the same lines as their real analogues. The covariance of complex white noise is defined by $\mathbb{E}\mathscr{W}(A)\overline{\mathscr{W}(B)} = \nu(A \cap B)$.

Here the variables $\mathscr{W}(A) \in \mathbf{C}$ and the integrands are complex-valued functions $f \in L_{2,\mathbf{C}}(\mathscr{R}, \mathscr{A}, \nu)$. Isometric property reads as

$$\mathbb{E}\left(\int_{\mathscr{R}} f \mathrm{d}\mathscr{W} \cdot \overline{\int_{\mathscr{R}} g \mathrm{d}\mathscr{W}}\right) = \int_{\mathscr{R}} f \overline{g} \mathrm{d}\nu.$$

Complex integration is necessary for spectral representation of a Gaussian stationary process

$$X(t) = \int_{-\infty}^{\infty} e^{itu} \mathrm{d}\mathscr{W}(u),$$

where the corresponding control measure ν is the spectral measure of process X. Even if X is a real-valued process, the corresponding noise $\mathscr{W}$ will be complex-valued.

3.2 Integral Representations

The properties of any centered Gaussian process $X(t), t \in T$, are entirely determined by its covariance function $K(s, t) = \mathbb{E}X(s)X(t)$. To construct such a process, it is enough to have a Gaussian white noise $\mathscr{W}$ on a measure space $(\mathscr{R}, \mathscr{A}, \nu)$ and a system of functions $\{m_t, t \in T\} \subset L_2(\mathscr{R}, \mathscr{A}, \nu)$ such that

$$(m_s, m_t)_2 = \int_{\mathscr{R}} m_s(u) m_t(u) \, \mathrm{d}\nu(u) = K(s, t), \quad s, t \in T.$$

In this case the process

$$\tilde{X}(t) = \int_{\mathscr{R}} m_t \mathrm{d}\mathscr{W}, \quad t \in T, \tag{3.1}$$

has a required covariance $K(s,t)$. We call expression (3.1) an *integral representation* of X.

Example 3.1 (*Wiener process*) We set $(\mathscr{R}, \mathscr{A}, \nu) = (\mathbb{R}_+, \mathscr{B}, \lambda)$, where $\mathscr{B}$ is Borel σ-field, λ is Lebesgue measure and

$$m_t(u) = \mathbf{1}_{[0,t]}(u). \tag{3.2}$$

Obviously,

$$\tilde{X}(t) = \int \mathbf{1}_{[0,t]}(u) \mathrm{d}\mathscr{W}(u) = \mathscr{W}([0, t])$$

is a Wiener process.

Example 3.2 (*Brownian bridge*) We give two representations for Brownian bridge W^0. The first one emerges from the representation for Wiener process through linear relation

$$W^0(t) = W(t) - t W(1). \tag{3.3}$$

Clearly, functions

$$m_t^0(u) = (m_t - t m_1)(u) = (1 - t)\mathbf{1}_{[0,t]}(u) - t\mathbf{1}_{(t,1]}(u)$$

provide an integral representation for W^0.

An alternative representation is built on a square $[0, 1]^2$ equipped with 2-dimensional Lebesgue measure. We let

$$\tilde{m}_t^0(u) = \mathbf{1}_{[0,t]\times[0,1-t]}(u).$$

Then (draw a picture!)

$$(\tilde{m}_s, \tilde{m}_t)_2 = \lambda^2 \left([0, s] \times [0, 1 - s] \bigcap [0, t] \times [0, 1 - t]\right)$$

$$= \min(s, t) \cdot \min(1 - s, 1 - t) = \min(s, t) \cdot (1 - \max(s, t))$$

$$= \min(s, t) - st,$$

as required for representation of W^0.

Example 3.3 (*fractional Brownian motion*) One can construct an integral representation for α-fractional Brownian motion $W^{(\alpha)}(t), t \in \mathbb{R}$ as follows [134]. Let $\mathscr{R} = \mathbb{R}$, $\nu = \lambda$ Lebesgue measure, and let $\mathscr{W}$ be a corresponding Gaussian white noise. Consider a process

$$W^{(\alpha)}(t) = \int_{-\infty}^{\infty} c_\alpha \left((t - u)^{\frac{\alpha-1}{2}} \mathbf{1}_{u \leq t} - (-u)^{\frac{\alpha-1}{2}} \mathbf{1}_{u \leq 0}\right) d\mathscr{W}(u). \tag{3.4}$$

Note that the integral is correctly defined exactly for $0 < \alpha < 2$. If the normalizing factor c_α is chosen appropriately, we obtain an α-fractional Brownian motion since for all $t \geq s$

$$\mathbb{E}\left(W^{(\alpha)}(t) - W^{(\alpha)}(s)\right)^2$$

$$= c_\alpha^2 \int_{-\infty}^{\infty} \left((t - u)^{\frac{\alpha-1}{2}} \mathbf{1}_{u \leq t} - (s - u)^{\frac{\alpha-1}{2}} \mathbf{1}_{u \leq s}\right)^2 du$$

$$\overset{u = s+v}{=} c_\alpha^2 \int_{-\infty}^{\infty} \left((t - s - v)^{\frac{\alpha-1}{2}} \mathbf{1}_{v \leq t-s} - (-v)^{\frac{\alpha-1}{2}} \mathbf{1}_{v \leq 0}\right)^2 dv$$

$$\overset{v = (t-s)w}{=} c_\alpha^2 (t - s)^\alpha \int_{-\infty}^{\infty} \left(1 - w)^{\frac{\alpha-1}{2}} \mathbf{1}_{w \leq 1} - (-w)^{\frac{\alpha-1}{2}} \mathbf{1}_{w \leq 0}\right)^2 dw$$

$$= const \cdot (t - s)^\alpha.$$

By computing the integral one can show that the required relation

$$\mathbb{E}\left(W^{(\alpha)}(t) - W^{(\alpha)}(s)\right)^2 = (t-s)^\alpha$$

is attained for

$$c_\alpha = \frac{\left[\sin(\frac{\pi\alpha}{2})\Gamma(\alpha+1)\right]^{1/2}}{\Gamma(\frac{\alpha+1}{2})}. \tag{3.5}$$

Exercise 3.2 (see [11, 108]) Prove the following useful relations between bifractional $W^{\alpha,K}$ and fractional $W^{(\alpha K/2)}$ Brownian motions.

$$W^{(\alpha K/2)}(t) = c_1 W^{\alpha,K}(t)$$
$$+ c_2 \int_0^\infty (1 - e^{-ut^\alpha})u^{-\frac{K+1}{2}} W(du), \; 0 < K < 1, 0 < \alpha < 2,$$

$$W^{\alpha,K}(t) = c_1 W^{(\alpha K/2)}(t)$$
$$+ c_2 \int_0^\infty (1 - e^{-ut^\alpha})u^{-\frac{K+1}{2}} W(du), \; 1 < K < 2, 0 < \alpha < \frac{2}{K},$$

where c_1 and c_2 are some appropriate positive constants depending on α, K and $W(du)$ denotes a standard white noise; the terms on the right hand side are independent.

Example 3.4 (*Riemann–Liouville processes and operators*) Opposite to previous and to subsequent examples, Riemann–Liouville process are defined via their integral representation and not via covariance.

Recall that Riemann–Liouville fractional integration operator is defined by

$$R_\alpha f(t) = \frac{1}{\Gamma(\alpha)} \int_0^t (t-u)^{\alpha-1} f(u) du, \quad \alpha > 0. \tag{3.6}$$

Here

$$R_\alpha : L_2[0, 1] \mapsto \begin{cases} L_p[0, 1], & \alpha > \frac{1}{2} - \frac{1}{p}, \\ \mathbb{C}[0, 1], & \alpha > \frac{1}{2}. \end{cases}$$

If $\alpha = 1$, we obtain the conventional integration operator. Riemann–Liouville operators have a remarkable semi-group property $R_\alpha R_\beta = R_{\alpha+\beta}$.

Similarly to (3.6), α-*Riemann–Liouville process* with $\alpha > 1/2$ is defined as a white noise integral on the real line (with Lebesgue measure as a control measure)

$$R^\alpha(t) = \frac{1}{\Gamma(\alpha)} \int_0^t (t-u)^{\alpha-1} d\mathscr{W}(u), \quad \alpha > 1/2.$$

When $\alpha = 1$, it coincides with a Wiener process. The restriction $\alpha > 1/2$ is necessary for correctness of the integral definition. In other words, the integration kernel $(t - \cdot)^{\alpha-1}$ must belong to $L_2[0, t]$.

The process R^α is H-self-similar with index $H = 2\alpha - 1$.

The semi-group property yields $R_\alpha R^\beta = R^{\alpha+\beta}$.

As one can observe from (3.4), for $\alpha \in (1/2, 3/2)$ the local properties of α-Riemann–Liouville process are close to those of α'-fractional Brownian motion with $\alpha' = 2\alpha - 1$.

However, opposite to fBm, the family of Riemann–Liouville processes has no limitations in sample path smoothness, because there is no upper bound for parameter α.

Unlike fBm, the process R^α is not a process with stationary increments. As a compensation, it has another property called extrapolation homogeneity [125]:

$$R^\alpha(t_0 + \cdot) - \mathbb{E}\left(R^\alpha(t_0 + \cdot)\big|\mathscr{F}_{t_0}\right) = R^\alpha(\cdot), \quad \forall t_0 \geq 0,$$

in distribution. Here $\mathscr{F}_{t_0}$ stands for the σ-field of the past for initial white noise $\mathscr{W}$ prior to time t_0, i.e.

$$\mathscr{F}_{t_0} = \sigma\{\mathscr{W}(A), A \subset [0, t_0]\}.$$

Remark 3.1 In econometric literature, where Riemann–Liouville process appears as a limit of discrete schemes, it is often called "fractional Brownian motion". Therefore, one should thoroughly avoid a confusion with the "true" fBm of Example 2.5. See [136] for further comparison of two processes.

Example 3.5 (*Brownian sheet*) This is a multi-parametric extension of Example 3.1. We let here $(\mathscr{R}, \mathscr{A}, v) = (\mathbb{R}_+^d, \mathscr{B}^d, \lambda^d)$, where $\mathscr{B}^d$ is Borel σ-field on $\mathbb{R}^d$, λ^d is d-dimensional Lebesgue measure, and define the "rectangles"

$$[0, t] := \{u \in \mathscr{R} : 0 \leq u_j \leq t_j, \ 1 \leq j \leq d\}.$$

Then the intersection of rectangles is again a rectangle (draw a picture):

$$[0, s] \cap [0, t] = \{u \in \mathscr{R} : 0 \leq u_j \leq \min(s_j, t_j), \ 1 \leq j \leq d\}.$$

Therefore, the functions

$$m_t(u) = \mathbf{1}_{[0,t]}(u)$$

have a property

$$(m_s, m_t)_2 = \lambda^d([0, s] \cap [0, t]) = \prod_{j=1}^{d} \min(s_j, t_j),$$

thus

$$\tilde{X}(t) = \int \mathbf{1}_{[0,t]}(u) \mathrm{d}\mathcal{W}(u) = \mathcal{W}([0,t])$$

is a Brownian sheet.

Example 3.6 (*Lévy Brownian function on $\mathbb{R}^d$*) Recall that LBf on $\mathbb{R}^d$ is defined by (2.7). We build now its white noise representation called *Chentsov integral-geometric construction* [37]. Let $\mathcal{R}$ be the space of all hyperplanes in $\mathbb{R}^d$. There exists a unique (up to a constant factor) measure v on $\mathcal{R}$ that is invariant with respect to all unitary transformations of $\mathbb{R}^d$. For $t \in \mathbb{R}^d$ let A_t denote the set of all hyperplanes crossing the segment $\overline{0,t} := \{rt, 0 \le r \le 1\}$. It is easy to observe that $v(A_t)$ is proportional to the length of $\overline{0,t}$. Let the measure v be normalized so that $v(A_t) = ||t||$. Let now $\mathcal{W}$ be a Gaussian white noise on $\mathcal{R}$ with control measure v. Then

$$W^L(t) := \mathcal{W}(A_t) = \int_{\mathcal{R}} \mathbf{1}_{A_t} \mathrm{d}\mathcal{W}, \quad t \in \mathbb{R}^d,$$

is a LBf on $\mathbb{R}^d$. Indeed, the relations $\mathbb{E}W^L(t) = 0$ and $W^L(0) = 0$ are obvious. Moreover, for any $s, t \in \mathbb{R}^d$ the set $A_s \Delta A_t$ consists of the hyperplanes crossing the segment $\overline{s,t} := \{s + rt, 0 \le r \le 1\}$ (we ignore the v-null set of hyperplanes that contain one of the points $0, s, t$). Therefore,

$$\mathbb{E}(W^L(s) - W^L(t))^2 = \mathbb{E}(\mathcal{W}(A_s) - \mathcal{W}(A_t))^2 = v(A_s \Delta A_t) = ||s - t||,$$

as required.

One may "pack" this integral-geometric construction into $\mathbb{R}^d$, rendering it more elementary although less transparent. Indeed, let $\mathcal{R}_0$ be the set of all hyperplanes crossing the origin. Then there exists a natural bijection between the sets $\mathcal{R} \backslash \mathcal{R}_0$ and $\mathbb{R}^d \backslash \{0\}$. i.e. to each hyperplane corresponds its point having the minimal distance to the origin. This bijection transforms the set A_t into the ball $\tilde{A}_t$ of radius $||t||/2$ centered at $\frac{t}{2}$ (the segment $\overline{0,t}$ is a diameter for this ball; check it!). The image of v is a spherically symmetric measure $\tilde{v} = \mathrm{d}r \, \mu(\mathrm{d}\theta)$, where $\mu(\mathrm{d}\theta)$ is appropriately normalized uniform measure on the unit sphere (prove it!). It is clear that the measure $\tilde{v}$ is different from Lebesgue measure on $\mathbb{R}^d$. If $\tilde{\mathcal{W}}$ is a white noise on $\mathbb{R}^d$ with control measure $\tilde{v}$, then $\tilde{\mathcal{W}}(\tilde{A}_t)$ is also a LBf.

One can implement an integral-geometric construction, similar to Chentsov construction, for LBf on a sphere [1], on a hyperbolic space, and in some other cases.

Example 3.7 (*Lévy Brownian function on L_1*, [118]). A centered Gaussian random function $\{W^L(t), t \in T\}$ parameterized by a metric space (T, ρ) with a marked point ϑ is called *Lévy Brownian function* (LBf) if $W^L(\vartheta) = 0$ and

$$\mathbb{E}(W^L(t) - W^L(s))^2 = \rho(s, t), \quad s, t \in T.$$

---

[1]   Instead of hyperplanes one should use the circles of maximal radius.

The spaces L_1 provide one of the most interesting space classes for which LBf is well defined. Indeed, let $T = L_1(U, \mathcal{U}, \mu)$ be our parametric space with a marked point $\vartheta = 0$, and with the L_1-distance

$$\rho(f, g) = \int_U |f(u) - g(u)| \mu(du).$$

Let $\mathcal{R} = U \times \mathbb{R}$, $\nu = \mu \times \lambda$ and consider a Gaussian white noise $\mathcal{W}$ on $\mathcal{R}$ with control measure ν. For any function $f \in T$ define its subgraph by

$$A_f = \{(u, r) : |r| \leq |f(u)|, rf(u) \geq 0, u \in U, r \in \mathbb{R}\} \subset \mathcal{R}.$$

Then

$$W^L(f) := \mathcal{W}(A_f) = \int_{\mathcal{R}} \mathbf{1}_{A_f} d\mathcal{W}, \quad f \in T,$$

is a LBf. Indeed,

$$\mathbb{E}(W_f^L - W_g^L)^2 = \mathbb{E}(\mathcal{W}(A_f) - \mathcal{W}(A_g))^2 = \int_{\mathcal{R}} (\mathbf{1}_{A_f} - \mathbf{1}_{A_g})^2 d\nu$$

$$= \nu(A_f \Delta A_g) = \int_U \mu(du) \lambda \left((A_f \Delta A_g) \cap (u \times \mathbb{R})\right)$$

$$= \int_U \mu(du) |f(u) - g(u)| = \|f - g\|_1 = \rho(f, g).$$

Exercise 3.3 Let S be the unit sphere in $\mathbb{R}^d$ and let μ be a uniform measure on S, normalized so that for any $x \in \mathbb{R}^d$ it is true that

$$\|x\| = \int_S |(x, u)| \mu(du).$$

Then $\mathbb{R}^d$ is isometrically embedded in $L_1(S, \mu)$ by

$$x \mapsto f_x(\cdot) := (x, \cdot).$$

By using this embedding, establish a connection between the integral representations of LBf on the spaces $\mathbb{R}^d$ and L_1, described in Examples 3.6 and 3.7, respectively.

Integral representations for stationary processes

We will consider a real-valued stationary Gaussian centered process $X(t), t \in \mathbb{R}$. By stationarity its covariance is $\mathbb{E}X(t)X(s) = K(t - s)$ where the function K is real-valued and non-negative definite. Therefore, K admits a spectral representation[2]

---

[2] We basically assume the basic spectral theory of stationary processes to be known and don't provide much details, see [182] for more.

$$K(\tau) = \int_{-\infty}^{\infty} e^{i\tau u} \nu(du), \quad \tau \in \mathbb{R},$$

where ν is a finite symmetric measure on $\mathbb{R}$. In particular, $\operatorname{Var} X(t) = \nu(\mathbb{R})$ for all t. The process X itself admits a spectral representation

$$X(t) = \int_{-\infty}^{\infty} e^{itu} \mathscr{W}(du), \quad t \in \mathbb{R}, \tag{3.7}$$

where $\mathscr{W}$ is a complex-valued Gaussian noise with uncorrelated but dependent values. One can express X via more conventional real-valued white noise as follows. Let $\mathscr{W}^{(re)}$, $\mathscr{W}^{(im)}$ be two independent copies of Gaussian white noise on $(0, \infty)$ controlled by the measure $\nu/2$ and let $\mathscr{W}_0$ be a centered Gaussian random variable with distribution $N(0, \nu(\{0\}))$ independent of $\mathscr{W}^{(re)}$, $\mathscr{W}^{(im)}$. Then $\mathscr{W}(\{0\}) = \mathscr{W}_0$ and

$$\mathscr{W}(A) = \begin{cases} \mathscr{W}^{(re)}(A) + i\mathscr{W}^{(im)}(A), & A \subset (0, \infty), \\ \mathscr{W}^{(re)}(-A) - i\mathscr{W}^{(im)}(-A), & A \subset (-\infty, 0). \end{cases}$$

Clearly, there is a dependence $\mathscr{W}(-A) = \overline{\mathscr{W}(A)}$ that actually provides the real values of X in the integral (3.7). In this case, our process writes as a sum of independent terms

$$X(t) = \mathscr{W}_0 + 2\int_0^{\infty} \cos(tu)\mathscr{W}^{(re)}(du) + 2\int_0^{\infty} \sin(tu)\mathscr{W}^{(im)}(du).$$

The just described construction admits various extensions, e.g. for random fields ($t, u \in \mathbb{R}^d$), random sequences ($t \in \mathbb{Z}, u \in \mathbb{S}^1$), and periodical processes ($t \in \mathbb{S}^1, u \in \mathbb{Z}$). It can be also extended to random processes and fields with stationary increments.

Let us now consider two more representations for stationary processes. Assume that the spectral measure ν has a density f and take any measurable function $\theta : \mathbb{R} \mapsto \mathbb{S}^1$. Then a family of functions $\{m_t, t \in \mathbb{R}\} \subset L_{2,\mathbb{C}}(\mathbb{R})$ is defined by

$$m_t(u) := \theta(u)e^{itu}\sqrt{f(u)}, \quad u \in \mathbb{R}.$$

Clearly,

$$\begin{aligned}
(m_t, m_s)_2 &= \int_{\mathbb{R}} m_t(u)\overline{m_s(u)}du \\
&= \int_{\mathbb{R}} |\theta(u)|^2 e^{i(t-s)u} f(u)du \\
&= \int_{\mathbb{R}} e^{i(t-s)u} \nu(du) \\
&= K(t, s) = \mathbb{E}X(t)X(s).
\end{aligned}$$

One can obtain more clear representation by applying to (m_t) the Fourier transform $\mathscr{F}: L_{2,\mathbb{C}}(\mathbb{R}) \mapsto L_{2,\mathbb{C}}(\mathbb{R})$. Let denote $h = \mathscr{F}\left(\theta(\cdot)\sqrt{f(\cdot)}\right)$ and consider the family

$$\tilde{m}_t(\cdot) = \mathscr{F}(m_t)(\cdot) = h(\cdot - t).$$

By isometric property of Fourier transform we have

$$(\tilde{m}_t, \tilde{m}_s)_2 = (m_t, m_s)_2 = \mathbb{E}X(t)X(s).$$

If an auxiliary function θ is chosen so cleverly that $h(\cdot)$ is real-valued, we obtain a *shift representation* of a process X via the white noise controlled by Lebesgue measure.

$$X(t) = \int_{\mathbb{R}} h(u - t)\mathscr{W}(du).$$

Example 3.8 (*Ornstein–Uhlenbeck process*) Recall that Ornstein–Uhlenbeck process defined in Example 2.6, has covariance function

$$K(\tau) = e^{-|\tau|/2} = \int_{\mathbb{R}} e^{i\tau u} \frac{2du}{\pi(1 + 4u^2)}.$$

Therefore, its spectral density is

$$f(u) = \frac{2}{\pi(1 + 4u^2)}.$$

By letting $\theta(u) = \frac{1+2iu}{(1+4u^2)^{1/2}}$ we obtain

$$m_t(u) = e^{itu}\left(\frac{2}{\pi(1 + 4u^2)}\right)^{1/2} \frac{1 + 2iu}{(1 + 4u^2)^{1/2}} = e^{itu}(2\pi)^{-1/2}(\frac{1}{2} - iu)^{-1}$$

and

$$h(v) = (\mathscr{F}m_0)(v) = e^{-v/2}\mathbf{1}_{v > 0}.$$

Therefore the shift representation of Ornstein–Uhlenbeck process has a form

$$X(t) = \int_t^\infty e^{-(v-t)/2}\mathscr{W}(dv).$$

4 Measurable Functionals and the Kernel

4.1 Main Definitions

After having collected a good deal of examples, we may continue construction of a general theory started in Sect. 1. We will still consider a Gaussian vector X taking values in a linear space $\mathscr{X}$ satisfying usual assumptions. Assume that $\mathbb{E}X = 0$ and let $K : \mathscr{X}^* \mapsto \mathscr{X}$ denote the covariance operator of X and $P = N(0, K)$ the distribution of X in $\mathscr{X}$.

Consider arbitrary continuous linear functional $f \in \mathscr{X}^*$. Since the random variable (f, x) is normally distributed, it has a finite second moment,

$$\mathbb{E}(f, X)^2 = \int_{\mathscr{X}} |f|^2 dP < \infty.$$

Therefore, a canonical embedding I^* of the dual space $\mathscr{X}^*$ into the Hilbert space $L_2(\mathscr{X}, P)$ is well defined. The closure of $I^*(\mathscr{X}^*)$ in $L_2(\mathscr{X}, P)$ is called the space of *measurable linear functionals* and denoted $\mathscr{X}_P^*$. It inherits from $L_2(\mathscr{X}, P)$ the scalar product

$$(z_1, z_2)_{\mathscr{X}_P^*} = (z_1, z_2)_2 = \int_{\mathscr{X}} z_1 z_2 dP = \mathbb{E}(z_1(X) z_2(X)).$$

In particular,

$$\|z\|_{\mathscr{X}_P^*}^2 = \mathbb{E}z(X)^2.$$

Every measurable linear functional is a limit (in $L_2(X, P)$ and P-almost surely) of a sequence of continuous linear functionals.

In the following it will be more convenient for us to consider I^* as an embedding

$$I^* : \mathscr{X}^* \mapsto \mathscr{X}_P^*.$$

The dual operator $I : \mathscr{X}_P^* \mapsto \mathscr{X}$ is defined by a natural relation

$$(f, Iz) = (I^* f, z)_{\mathscr{X}_P^*} = \mathbb{E}(f, X) z(X), \quad \forall f \in \mathscr{X}^*, z \in \mathscr{X}_P^*.$$

The existence of this dual operator is not obvious, since we said nothing about the continuity of the initial operator I^*. It is, however possible to prove that under usual assumptions stated in Sect. 1 the dual operator indeed exists.

The operator I is linear and injective: if $Iz = 0$ holds for some $z \in \mathscr{X}_P^*$, then we have

$$(I^* f, z) = (f, Iz) = 0$$

for any $f \in \mathscr{X}^*$. By considering a sequence $I^* f_n$ converging to z, we obtain $\|z\|_{\mathscr{X}_P^*} = 0$, i.e. $z = 0$.

It is also important to notice that covariance operator admits a factorization

$$K = II^*. \tag{4.1}$$

Indeed,

$$(f, II^*g) = (I^*f, I^*g) = \mathbb{E}(f, X)(g, X) = (f, Kg) \quad \forall f, g \in \mathscr{X}^*.$$

Now we are able to give a fundamental definition: the set $H_P := I(\mathscr{X}_P^*) \subset \mathscr{X}$ equipped with the scalar product

$$(h_1, h_2)_{H_P} := \left(I^{-1}h_1, I^{-1}h_2 \right)_{\mathscr{X}_P^*}, \quad h_1, h_2 \in H_P,$$

and with the corresponding norm

$$|h|_{H_P}^2 := (h, h)_{H_P}, \quad h \in H_P,$$

is called the *kernel* of distribution P.

Correctness of the norm definition is guaranteed by the injection property of operator I. The unit ball $\{h \in H_P : |h|_{H_P} \leq 1\}$ is sometimes called *dispersion ellipsoid* of measure P. We can also call H the kernel of vector X but in this case one should remember that the kernel is entirely determined by the vector's distribution.

The following exposition will demonstrate that the kernel contains all important information about P and X; the solution to any important problem is expressed in terms of the kernel. Let us first mention some simple properties of the kernel.

- By (4.1) we have $K(\mathscr{X}^*) \subset H_P \subset \mathscr{X}$. If H_P is finite-dimensional, then all three spaces coincide. Otherwise, they are all different.
- If H_P is infinite-dimensional, then $P(H_P) = 0$ (in spite of all importance H_P for description of P!)
- Topological support of measure P coincides with the closure of H_P in $\mathscr{X}$.
- The space H_P is separable.
- The balls $\{h \in H_P : |h|_{H_P} \leq r\}$ are compact sets in $\mathscr{X}$.

Example 4.1 (*Standard Gaussian measure in* $\mathbb{R}^\infty$). Consider a standard Gaussian vector X defined in Example 2.1. For any $f, g \in \mathbf{c}_0 = \mathscr{X}^*$ we have

$$(I^*f, I^*g)_{X_P^*} = \mathbb{E}\left(\sum_{j=1}^\infty f_j X_j \right)\left(\sum_{j=1}^\infty g_j X_j \right) = \sum_{j=1}^\infty f_j g_j = (f, g)_2.$$

It follows that a generic form of a measurable linear functional $z \in \mathscr{X}_P^*$ is

$$z(x) = \sum_{j=1}^\infty z_j x_j, \quad (z_j) \in \ell_2.$$

Moreover, by using coordinate functionals δ_k we can write

$$(Iz)_k = (\delta_k, Iz) = (I^*\delta_k, z)_{\mathscr{X}_P^*} = \mathbb{E}\left[X_k \left(\sum_{j=1}^\infty z_j X_j \right) \right] = z_k.$$

This means that $H_P = \ell_2$, $(h_1, h_2)_{H_P} = (h_1, h_2)_2$ and $|h|_{H_P} = ||h||_2$.

Example 4.2 (*Gaussian vectors in a Hilbert space*) Consider a Gaussian vector X in a Hilbert space defined in Example 2.2. For any $f, g \in \mathscr{X} = \mathscr{X}^*$ we have

$$(I^* f, I^* g)_{X_P^*} = \mathbb{E} \left(\sum_{j=1}^{\infty} f_j \sigma_j \xi_j \right) \left(\sum_{j=1}^{\infty} g_j \sigma_j \xi_j \right) = \sum_{j=1}^{\infty} f_j g_j \sigma_j^2.$$

It follows that

$$\mathscr{X}_P^* = \left\{ z : z(x) = \sum_{j=1}^{\infty} z_j x_j, \ \sum_{j=1}^{\infty} z_j^2 \sigma_j^2 < \infty \right\},$$

$$||z||_{\mathscr{X}_P^*}^2 = \sum_{j=1}^{\infty} z_j^2 \sigma_j^2, \quad z \in \mathscr{X}_P^*.$$

Again, by using coordinate functionals, we obtain

$$(e_k, Iz) = (I^* e_k, z) = \mathbb{E} \left(\sigma_k \xi_k \left(\sum_{j=1}^{\infty} z_j \sigma_j \xi_j \right) \right) = \sigma_k^2 z_k,$$

i.e.

$$Iz = \sum_{j=1}^{\infty} \sigma_j^2 z_j e_j.$$

If $h = Iz \in H_P$, then

$$||h||_{H_P}^2 = ||z||_{X_P}^2 = \sum_{j=1}^{\infty} z_j^2 \sigma_j^2 = \sum_{j=1}^{\infty} \frac{h_j^2}{\sigma_j^2}.$$

This means that

$$H_P = \left\{ h \in \mathscr{X} : \sum_{j=1}^{\infty} \frac{h_j^2}{\sigma_j^2} < \infty \right\}$$

and

$$(h_1, h_2)_{H_P} = \sum_{j=1}^{\infty} \frac{(h_1)_j (h_2)_j}{\sigma_j^2},$$

where $h_j = (h, e_j)$ are the coordinates of h in the base (e_j).

Example 4.3 (*Finite-dimensional Gaussian vector*) Let $\mathcal{X} = \mathbb{R}^n$ and assume that a vector X has a distribution $P = N(0, K)$ with a non-degenerate covariance operator $K : \mathbb{R}^n \mapsto \mathbb{R}^n$, i.e. $K(\mathbb{R}^n) = \mathbb{R}^n$. Then K is a diagonal operator with respect to some base (e_j),

$$K\left(\sum_{j=1}^{n} x_j e_j\right) = \sum_{j=1}^{n} \sigma_j^2 x_j e_j,$$

where $\sigma_j > 0$ for all j. It follows that X can be represented by $X = \sum_{j=1}^{n} \sigma_j \xi_j e_j$, as considered in the previous example. By applying the results we obtained there, we see that $H_P = \mathbb{R}^n$ and

$$(h_1, h_2)_{H_P} = \sum_{j=1}^{\infty} \frac{(h_1)_j (h_2)_j}{\sigma_j^2} = (K^{-1} h_1, h_2),$$

$$|h|_{H_P}^2 = (K^{-1} h, h) = ||K^{-1/2} h||^2.$$

4.2 Factorization Theorem

Unlike the few mentioned examples, in most cases there is no elementary trick for immediate determination of the kernel for a Gaussian distribution. For this purpose, the following result will be useful.

Theorem 4.1 (Factorization Theorem) *Let $\mathcal{H}$ be a Hilbert space and let $J : \mathcal{H} \mapsto \mathcal{X}$ be an injective linear mapping such that factorization*

$$K = J J^*$$

holds. Then the kernel of P can be expressed as $H_P = J(\mathcal{H})$, while the scalar product and the norm in H_P admit representations

$$(h_1, h_2)_{H_P} = (J^{-1} h_1, J^{-1} h_2)_{\mathcal{H}}, \quad \forall h_1, h_2 \in H_P,$$

$$|h|_{H_P} = ||J^{-1} h||_{\mathcal{H}}, \quad \forall h \in H_P. \tag{4.2}$$

Remark 4.1 If the operator J is not injective, the equality $H_P = J(\mathcal{H})$ still holds but (4.2) should be replaced by

$$|h|_{H_P} = \inf_{\ell : J\ell = h} \{||\ell||_{\mathcal{H}}\}, \quad \forall h \in H_P.$$

Remark 4.2 We have already shown that operator I generates a required factorization and can be formally considered within theorem's frame. The main advantage of

Factorization Theorem is, however, the freedom of choice of $\mathcal{H}$ and J for calculation convenience.

Proof (*of the theorem*). Let us define an isometry U between the spaces $I^*(\mathcal{X}^*) \subset X_P^*$ and $J^*(\mathcal{X}^*) \subset \mathcal{H}$ by the relation

$$UI^*f := J^*f, \quad f \in \mathcal{X}^*.$$

It is true that

$$(UI^*f, UI^*g)_{\mathcal{H}} = (J^*f, J^*g)_{\mathcal{H}} = (f, JJ^*g) = (f, Kg) = (f, II^*g)$$
$$= (I^*f, I^*g)_{\mathcal{X}_P^*}.$$

and we see that the scalar product (hence, norms and distances) is conserved by U. Clearly, isometry U can be extended to the respective closures. By the definition, the closure of $I^*(\mathcal{X}^*)$ is X_P^*. Moreover, we can prove that the closure of $J^*(\mathcal{X}^*)$ coincides with $\mathcal{H}$. Indeed, if an element $\ell \in \mathcal{H}$ is orthogonal to $J^*(\mathcal{X}^*)$, then for all $f \in \mathcal{X}^*$

$$0 = (J^*f, \ell)_{\mathcal{H}} = (f, J\ell).$$

It follows that $J\ell = 0$ and injective property of J yields $\ell = 0$. Therefore, the extension of U is an isometry of spaces X_P^* and $\mathcal{H}$. Let us check an operator identity

$$I = JU. \tag{4.3}$$

Assuming this is proved, we have

$$H_P = I(X_P^*) = JU(X_P^*) = J(\mathcal{H}),$$

$$|h|_{H_P} = ||I^{-1}h||_{X_P^*} = ||U^{-1}J^{-1}h||_{X_P^*} = ||J^{-1}h||_{\mathcal{H}}.$$

Thus, let us prove (4.3). For any $z \in X_P^*$, $f \in \mathcal{X}^*$ we have

$$(f, JUz) = (J^*f, Uz)_{\mathcal{H}} = (U^{-1}J^*f, z)_{X_P^*} = (I^*f, z) = (f, Iz).$$

It follows that $(JU)z = Iz$. □

Example 4.4 (*Wiener process*) We find the kernel for the distribution of Wiener process defined in Example 2.3. Let $\mathcal{X} = \mathbb{C}[0, 1]$, $X = W$, $\mathcal{H} = L_2[0, 1]$. We define operator $J : \mathcal{H} \mapsto \mathcal{X}$ as integration operator

$$(J\ell)(t) = \int_0^t \ell(s)ds.$$

Recall that $\mathcal{X}^* = \mathbb{M}[0, 1]$ is a space of charges (sign measures) on $[0, 1]$. It is easy to check that $J^* : \mathbb{M}[0, 1] \mapsto \mathcal{H}$ is given by

$$(J^*\mu)(s) = \mu[s, 1].$$

It follows that

$$
\begin{aligned}
(JJ^*\mu)(t) &= \int_0^t (J^*\mu)(s)\mathrm{d}s = \int_0^t \mu[s, 1]\mathrm{d}s \\
&= \int_0^1 \int_0^1 \mathbf{1}_{s\le t}\mathbf{1}_{s\le u}\mu(\mathrm{d}u)\mathrm{d}s \\
&= \int_0^1 \min\{t, u\}\mu(\mathrm{d}u) = K\mu(t).
\end{aligned}
$$

Therefore, factorization assumption is verified. Moreover, J is an injective operator. By applying Theorem 4.1, we conclude that the kernel of Wiener measure is given by

$$
\begin{aligned}
H_P &= \left\{ h : h(t) = \int_0^t \ell(s)\mathrm{d}s, \ell \in L_2[0, 1] \right\} \\
&= \{ h : h \in AC[0, 1], h(0) = 0, h' \in L_2[0, 1] \}.
\end{aligned}
\tag{4.4}
$$

Here $AC[0, 1]$ denotes the class of absolutely continuous functions. The norm and the scalar product in H_P are given by

$$|h|^2_{H_P} = \int_0^1 h'(s)^2 \mathrm{d}s,$$

$$(h_1, h_2)_{H_P} = \int_0^1 h'_1(s)h'_2(s)\mathrm{d}s.$$

This is actually a Sobolev space W_2^1 with a one-sided boundary condition. The kernel of Wiener measure was discovered in the works of Cameron and Martin [34] and was the first investigated kernel. Therefore, it is often called *Cameron–Martin space*. Sometimes this name is even extrapolated to the kernels of other Gaussian distributions.

Remark 4.3 The kernel (considered as a set and as a Hilbert space) actually does not depend on the space in which we consider our Gaussian vector. For example, if a Wiener process is considered as a random element of the space $\tilde{\mathscr{X}} = L_2[0, 1]$, the kernel turns out to be the same Cameron–Martin space.

Example 4.5 (The kernel of a process admitting an integral representation). Consider a Gaussian process $X(t), t \in T$, admitting an integral representation with respect to a white noise on a space $(\mathscr{R}, \mathscr{A}, \nu)$,

$$X(t) = \int_{\mathscr{R}} m_t(u)\mathscr{W}(\mathrm{d}u), \quad t \in T, m_t \in L_2(\mathscr{R}, \mathscr{A}, \nu).$$

which is equivalent to representation of covariance function of X in the form

$$K(s,t) := \mathbb{E}X(s)X(t) = \int_{\mathscr{R}} m_s m_t dv, \quad s, t \in T.$$

In order to avoid unnecessary topological details, let us assume that T is a compact topological space and that the sample paths of X are continuous on T. Then we may assume that X is a random vector in $\mathscr{X} = \mathbb{C}(T)$ having integral covariance operator (cf. (2.2), Example 2.4).

Let $\mathscr{H} = L_2(\mathscr{R}, \mathscr{A}, v)$ and define an operator $J : \mathscr{H} \mapsto \mathscr{X}$ by the formula

$$(J\ell)(t) := (\ell, m_t)_{\mathscr{H}} = \int_{\mathscr{R}} \ell m_t dv, \quad t \in T.$$

The dual operator $J^* : \mathbb{M}(T) \mapsto \mathscr{H}$ writes as

$$(J^*\mu)(u) = \int_T m_s(u)\mu(ds)$$

and we obtain

$$(JJ^*\mu)(t) = \int_{\mathscr{R}} (J^*\mu)(u)m_t(u)v(du) = \int_{\mathscr{R}} \int_T m_s(u)\mu(ds)m_t(u)v(du)$$

$$= \int_T \int_{\mathscr{R}} m_s(u)m_t(u)v(du)\mu(ds)$$

$$= \int_T K(s,t)\mu(ds) = (K\mu)(t).$$

Therefore, factorization assumption is verified. By applying Theorem 4.1, we see that

$$H_P = \left\{ h : h(t) = \int_{\mathscr{R}} \ell(u)m_t(u)v(du), \ell \in L_2(\mathscr{R}, \mathscr{A}, v) \right\}. \tag{4.5}$$

If the operator J is injective, we find the norm

$$|h|_P^2 = \int l(u)^2 v(du)$$

and the respective scalar product.

The kernel of Wiener process (4.4) can be obtained as a special case of this construction by using representation (3.2). We will now discuss two more examples.

Example 4.6 (Kernel of a fractional Brownian motion [77]). Recall that α-fractional Brownian motion (fBm) $W^{(\alpha)}(t)$, $t \in \mathbb{R}$, was defined in Example 2.5 and its integral representation was calculated in Example 4.6. By plugging this representation in (4.5), we find the kernel

$$\left\{ h : h(t) = \int_{-\infty}^{\infty} c_\alpha \left((t-u)^{\frac{\alpha-1}{2}} 1_{u \leq t} - (-u)^{\frac{\alpha-1}{2}} 1_{u \leq 0} \right) \ell(u) du, \quad \ell \in L_2(\mathbb{R}) \right\}.$$

In other words, a function belongs to the kernel of fBm, if its fractional derivative of order $\frac{\alpha+1}{2}$ is square integrable.

Exercise 4.1 Find the kernel of α-Riemann–Liouville process defined in Example 3.4.

Example 4.7 (Kernel of Brownian sheet) Recall that Brownian sheet $W(t), t \in \mathbb{R}^d_+$ was defined in Example 2.7 and its integral representation was found in Example 3.5. By plugging this representation in (4.5) we find the kernel

$$\left\{ h : h(t) = \int_{[0,t]} \ell(u) \lambda^d (du), \quad \ell \in L_2(\mathbb{R}^d_+) \right\}.$$

in other words, a function $h : \mathbb{R}^d_+ \mapsto \mathbb{R}$ belongs to the kernel of Brownian sheet if its mixed derivative $\frac{\partial^d}{\partial t_1 \ldots \partial t_d} h$ is square integrable.

It will be useful to study the behavior of kernels under linear transformations. Let $X \in \mathcal{X}$ be a centered Gaussian vector with covariance operator K_X and distribution $P = N(0, K_X)$, and let $L : \mathcal{X} \to \mathcal{Y}$ be a continuous linear mapping. Consider $Y := LX$, the image of X under L. Obviously, Y is a Gaussian vector in $\mathcal{Y}$ with covariance operator $K_Y = L K_X L^*$ and distribution $Q := PL^{-1}$.

Proposition 4.1 *It is true that*

$$H_Q = L(H_P)$$

and

$$|v|_{H_Q} = \inf_{h \in L^{-1} v} |h|_{H_P}, \quad \forall v \in H_Q.$$

Proof. Let I, I^* denote the canonical operators related to vector X. Then we have a factorization

$$K_Y = L K_X L^* = L I I^* L^* = (LI)(LI)^*.$$

By Theorem 4.1 we obtain $H_Q = (LI)(\mathcal{X}^*_P) = L(H_P)$, and for any $v \in H_Q$

$$|v|_{H_Q} = \inf_{z \in (LI)^{-1} v} ||z||_{\mathcal{X}^*_P} = \inf_{z \in (LI)^{-1} v} |Iz|_{H_P} = \inf_{h \in L^{-1} v} |h|_{H_P}.$$

$\square$

We demonstrate this Proposition's action by two examples.

Example 4.8 (Kernel of Ornstein–Uhlenbeck process) Recall that Ornstein–Uhlenbeck process was defined in Example 2.6 as a linear transformation of Wiener process, $X(t) = e^{-t/2} W(e^t)$. Let $\mathcal{X} = \mathbb{C}[0, e]$, $\mathcal{Y} = \mathbb{C}[0, 1]$, and define a linear operator L by $(Lw)(t) = e^{-t/2} w(e^t)$. According to Proposition 4.1, we have to find the L-image of Wiener kernel (4.4). Let $v = Lh$. Then h admits a partial expression via v by

$$h(s) = \sqrt{s}v(\ln s), \quad 1 \le s \le e.$$

Hence, $h(1) = v(0)$ and

$$
\begin{aligned}
\int_1^e h'(s)^2 ds &= \int_1^e \left(\frac{v(\ln s)}{2\sqrt{s}} + \frac{\sqrt{s}v'(\ln s)}{s} \right)^2 ds \\
&= \int_1^e \left(\frac{v(\ln s)}{2} + v'(\ln s) \right)^2 \frac{ds}{s} \\
&= \int_0^1 \left(\frac{v(t)}{2} + v'(t) \right)^2 dt \\
&= \frac{1}{4} \int_0^1 v(t)^2 dt + \int_0^1 v(t)v'(t) dt + \int_0^1 v'(t)^2 dt \\
&= \frac{1}{4} \int_0^1 v(t)^2 dt + \frac{v(1)^2 - v(0)^2}{2} + \int_0^1 v'(t)^2 dt.
\end{aligned}
$$

The function h is not uniquely defined on the interval $[0, 1]$ by the relation; we only know that $h(0) = 0$, $h(1) = v(0)$. Under these conditions the minimum of the integral of squared derivative is attained on a linear function and is equal to $v(0)^2$, since

$$\int_0^1 h'(s)^2 ds \ge \left(\int_0^1 h'(s) ds \right)^2 = (h(1) - h(0))^2 = h(1)^2 = v(0)^2.$$

We conclude that the kernel H^U of Ornstein–Uhlenbeck process in $\mathbb{C}[0, 1]$ has a form

$$H^U = \{v : v \in AC[0, 1], \; v' \in L_2[0, 1]\}$$

and

$$|v|_{H^U}^2 = \frac{v(1)^2 + v(0)^2}{2} + \frac{1}{4} \int_0^1 v(t)^2 dt + \int_0^1 v'(t)^2 dt, \quad v \in H^U.$$

The stationarity of the process is reflected by the invariance of the norm with respect to the shift and to the inversion of time.

Exercise 4.2   Find the kernel of Ornstein–Uhlenbeck process in $\mathbb{C}[a, b]$ for arbitrary interval $[a, b] \subset \mathbb{R}$. Solve analogous problem for stationary Gaussian process X with covariance function having a slightly more general form $\mathbb{E}X(s)X(t) = \alpha e^{-\beta|s-t|}$.

Example 4.9   (*Kernel of Brownian bridge*) Recall that one of possible definitions of Brownian bridge is

$$W^0(t) = W(t) - tW(1), \quad 0 \le t \le 1,$$

where W is a Wiener process. This means that $W^0 = LW$, where operator L : $\mathbb{C}[0, 1] \mapsto \mathbb{C}[0, 1]$ is given by $(Lw)(t) = w(t) - tw(1)$. It follows that the kernel of Brownian bridge H^0 is the L-image of the kernel of Wiener process (4.4). It is easy to establish that

$$H^0 = \left\{ v : v \in AC[0, 1], v(0) = v(1) = 0, \; v' \in L_2[0, 1] \right\}.$$

Indeed, any function on the right-hand side of equality belongs to Wiener kernel and satisfies equation $Lh = h$. Therefore, it belongs to H^0. Conversely, if a function h belongs to the L-image of Wiener kernel, then it satisfies the conditions mentioned on the equality's right-hand side.

It remains to reconstruct the Hilbert structure in the kernel. Let $v \in H^0$ and $Lh = v$. Then $h(t) = v(t) + th(1)$, hence $h'(t) = v'(t) + h(1)$ and

$$
\begin{aligned}
|h|^2 &= \int_0^1 h'(t)^2 dt = \int_0^1 v'(t)^2 dt + h(1)^2 + 2h(1) \int_0^1 v'(t) dt \\
&= \int_0^1 v'(t)^2 dt + h(1)^2,
\end{aligned}
$$

since

$$\int_0^1 v'(t) dt = v(1) - v(0) = 0.$$

The minimum of this expression over $h \in L^{-1}v$ is attained when $h(1) = 0$, i.e. $h = v$. We conclude that

$$|h|_{H^0}^2 = \int_0^1 h'(t)^2 dt,$$

that is the norm and the scalar product in H^0 are the same as in the kernel of the Wiener measure.

4.3 Alternative Approaches to Kernel Definition

Reproducing kernel (RKHS)

A popular alternative approach to kernel's construction is known as a concept of *Reproducing Kernel* or *Reproducing Kernel Hilbert Space, RKHS* [5]. Let T be an arbitrary set, and let $K : T \times T \to \mathbb{R}$ be a non-negative definite function (called a kernel). The space H reproducing the kernel K is a class of functions $f : T \mapsto \mathbb{R}$. It is constructed as follows. We take a linear span of functions $K(t, \cdot), t \in T$, and introduce a scalar product by

$$\langle K(s, \cdot), K(t, \cdot)\rangle := K(s, t), \quad s, t \in T.$$

Then H is obtained by a completion of this span with respect to the Hilbert distance.

Let us establish the link between this construction and our notion of the kernel. Let T be a compact metric space, let $X \in \mathbb{C}(T)$ be a centered Gaussian random process with continuous sample paths and covariance function $K(s, t)$. For any $t \in T$ let us consider a functional $(\delta_t, x) := x(t)$ and let $h_t := II^*\delta_t$. Then

$$h_t(s) = (\delta_s, II^*\delta_t) = (I^*\delta_s, I^*\delta_t)_{\mathscr{X}_P^*} = \mathbb{E}X(s)X(t) = K(s, t),$$

i.e. $h_t = K(\cdot, t)$. The scalar product of two elements of this form is the same as above, i.e.

$$(h_s, h_t)_{H_P} = (II^*\delta_s, II^*\delta_t)_{H_P} = (I^*\delta_s, I^*\delta_t)_{\mathscr{X}_P^*} = K(s, t).$$

It follows that reproducing kernel is a closed subset of H_P. By approximating arbitrary linear functionals by linear combinations of δ-functionals, it is not hard to establish that two spaces actually coincide.

Vector integrals

Assume that we are able to define vector-valued integrals

$$\mathbb{E}z(X)X = \int_{\mathscr{X}} z(x)x P(\mathrm{d}x), \quad z \in \mathscr{X}_P^*,$$

correctly. Then

$$(f, \mathbb{E}z(X)X) = \mathbb{E}z(X)(f, X) = (z, I^*f)_{\mathscr{X}_P^*} = (f, Iz), \quad \forall z \in \mathscr{X}_P^*, f \in \mathscr{X}^*.$$

We obtain a representation of the kernel via vector-valued integrals [26, 105]

$$H_P = \{Iz, z \in \mathscr{X}_P^*\} = \left\{\int_{\mathscr{X}} z(x)x P(\mathrm{d}x), z \in \mathscr{X}_P^*\right\}.$$

Polar representation

Consider a convex set of continuous linear functionals

$$B = \left\{f \in \mathscr{X}^* : \mathbb{E}(f, X)^2 = \int_{\mathscr{X}} (f, x)^2 P(\mathrm{d}x) = ||I^*f||_{\mathscr{X}_P^*}^2 \le 1\right\}.$$

For any $h = Iz \in H_P$ we have

$$\sup_{f \in B} |(f, h)| = \sup_{f \in B} |(f, Iz)| = \sup_{f \in B} |(I^*f, z)_{\mathscr{X}_P^*}| \le ||z||_{\mathscr{X}_P^*} = |h|_{H_P}.$$

By approximating z by continuous functionals, it is easy to see that actually we have an equality and

$$H_P = \left\{h \in \mathscr{X} : \sup_{f \in B} |(f, h)| < \infty\right\}.$$

5 Cameron–Martin Theorem

Let X be a random vector taking values in a linear space $\mathscr{X}$, let P be the distribution of X, and let h be a vector in $\mathscr{X}$; then the distribution P_h of the vector $X + h$ is defined by formula

$$P_h(A) = P(A - h), \quad \forall A \subset \mathscr{X},$$

and is called a *shift of P in direction h*. We are interested in checking the absolute continuity of P_h with respect to P. Recall that a measure Q is *absolutely continuous with respect to* P (we write $Q \ll P$), if $P(A) = 0$ yields $Q(A) = 0$. There is an equivalent property: there exists a *density* $g := \frac{dQ}{dP} \in L_1(\mathscr{X}, P)$ such that

$$Q(A) = \int_A g \, dP, \quad \forall A \subset \mathscr{X}.$$

If $P_h \ll P$, then h is called *admissible shift* for P. If ch is an admissible shift for P for all $c \in \mathbb{R}$ then we say that h defines an *admissible direction* for P.

We are interested in the case of Gaussian measures, i.e. $P = N(a, K)$ and $P_h = N(a + h, K)$. One can essentially observe what is going on even on the real line.

Example 5.1 Let $\mathscr{X} = \mathbb{R}$ and $P = N(0, 1)$, $h \in \mathbb{R}$. Then the density of P with respect to Lebesgue measure is

$$p(x) = \frac{1}{\sqrt{2\pi}} \exp\{-x^2/2\},$$

while for the measure $P_h = N(h, 1)$ we have a density

$$p_h(x) = \frac{1}{\sqrt{2\pi}} \exp\{-(x - h)^2/2\}.$$

Therefore $P_h \ll P$ and

$$\frac{dP_h}{dP}(x) = \frac{p_h(x)}{p(x)} = \exp\left\{hx - \frac{h^2}{2}\right\}.$$

In general case, not every shift is an admissible one but for admissible shifts the form of the density is exactly the same: it is an exponent of linear functional multiplied by a normalizing quadratic constant.

Theorem 5.1 (Cameron–Martin theorem) *Let P be a centered Gaussian measure in a linear space $\mathscr{X}$. Then $P_h \ll P$ iff $h \in H_P$. If $h \in H_P$, then the density $\frac{dP_h}{dP}$ has a form*

$$\frac{dP_h}{dP}(x) = \exp\left\{z(x) - \frac{|h|^2_{H_P}}{2}\right\}. \tag{5.1}$$

where $z \in \mathscr{X}^*_P$ is a measurable linear functional such that $Iz = h$.

It follows from the theorem that for Gaussian measure every admissible shift defines an admissible direction. The density formula (5.1) is called *Cameron–Martin formula*. More precisely, Cameron and Martin considered (in late forties of XX century) not the general case but a Wiener process for which we obtain the following corollary. Let $\mathscr{X} = \mathbb{C}[0, 1]$, let $X = W$ be a Wiener process, let P and H_P be its distribution and kernel, respectively. Recall that

$$H_P = \{h \in AC[0, 1], h(0) = 0, h' \in L_2[0, 1]\},$$

$$|h|^2_{H_P} = \int_0^1 h'(s)^2 ds, \quad h \in H_P.$$

As for the functional z associated to vector h by formula $Iz = h$, it is easy to show that it coincides with Wiener integral

$$z(w) = \int_0^1 h'(s)dw(s).$$

Indeed, by taking into account the isometric property of Wiener integral, for any $t \in [0, 1]$ we obtain

$$Iz(t) = \delta_t(Iz) = (I^*\delta_t, z)$$

$$= \mathbb{E}W(t)z(W) = \left(\int_0^1 \mathbf{1}_{[0,t]}(s)dw(s) \cdot \int_0^1 h'(s)dw(s)\right)$$

$$= \int_0^t h'(s)ds = h(t).$$

Therefore, Cameron–Martin formula for Wiener process reads as

$$\frac{dP_h}{dP}(w) = \exp\left\{\int_0^1 h'(s)dw(s) - \frac{1}{2}\int_0^1 h'(s)^2 ds\right\}.$$

Proof (of Cameron–Martin theorem).

1) *Sufficiency.* Let $h \in H_P$. Consider a measure

$$Q(dx) = \exp\left\{z(x) - \frac{|h|^2_{H_P}}{2}\right\} P(dx).$$

It is sufficient for us to show that $Q = P_h$. We will check that the characteristic functions of two measures coincide. Let $f \in \mathscr{X}^*$. Then

$$\int_{\mathscr{X}} e^{i(f,x)} P_h(dx) = \mathbb{E}e^{i(f,X+h)} = e^{i(f,h)-(f,Kf)/2},$$

where X is a vector with distribution P. On the other hand,

$$\int_{\mathscr{X}} e^{i(f,x)} Q(dx) = \int_{\mathscr{X}} \exp\left\{i(f,x) + z(x) - \frac{|h|^2_{H_P}}{2}\right\} P(dx)$$

$$= \exp\left\{-\frac{|h|^2_{H_P}}{2}\right\} \mathbb{E}\exp\{i(f,X) + z(X)\}$$

Recall that the two-dimensional vector $Y = ((f,X), z(X))$ is a centered Gaussian vector with covariance matrix

$$K^Y = \begin{pmatrix} (f,Kf) & (f,h) \\ (f,h) & |h|^2_{H_P} \end{pmatrix}.$$

Indeed, $\mathbb{E}(f,X)^2 = (f,Kf)$ by the definition of covariance operator,

$$\mathbb{E}(f,X)z(X) = (I^*f, z)_{\mathscr{X}^*_P} = (f, Iz) = (f,h)$$

and

$$\mathbb{E}z(X)^2 = ||z||^2_{\mathscr{X}^*_P} = |h|^2_{H_P}.$$

The well-known formula for finite-dimensional Gaussian vectors yields

$$\mathbb{E}e^{v_1 Y_1 + v_2 Y_2} = \left\{\frac{K^Y_{11}v_1^2 + 2K^Y_{12}v_1 v_2 + K^Y_{22}v_2^2}{2}\right\}.$$

By plugging $v_1 = i$, $v_2 = 1$ in this formula, we obtain

$$\mathbb{E}\exp\{i(f,X) + z(X)\} = \exp\left\{\frac{-(f,Kf) + 2i(f,h) + |h|^2_{H_P}}{2}\right\}.$$

Therefore,

$$\int_{\mathscr{X}} e^{i(f,x)} Q(dx) = \exp\left\{\frac{-(f,Kf) + 2i(f,h)}{2}\right\},$$

and we see that the characteristic functions are the same.

2) *Necessity.* Let h be an admissible shift. We show that $h \in H_P$. For achieving this goal, we have to find a measurable linear functional $z \in \mathscr{X}^*_P$ such that $h = Iz$. For every $f \in \mathscr{X}^*$ let μ_f and ν_f denote the distributions of random variables (f, X) and $(f, X+h)$, respectively. These are normal distributions: if $P = N(0, K)$, then $\mu_f = N(0, (f, Kf))$, $\nu_f = N((f,h), (f, Kf))$.
Since $P_h \ll P$, for any $\varepsilon > 0$ there exists $\delta > 0$ such that for any measurable $A \subset \mathscr{X}$, the inequality $P(A) \le \delta$ yields $P_h(A) \le \varepsilon$. In particular, for any Borel set $B \subset \mathbb{R}$, the inequality $\mu_f(B) \le \delta$ yields $\nu_f(B) \le \varepsilon$.

The stated property is possible only if the difference of barycenters of the distributions μ_f and ν_f is commeasurable with with their common quadratic deviation $(f, Kf)^{-1/2}$:

$$\sup_{f \in \mathscr{X}^*, f \neq 0} \frac{|(f, h)|}{(f, Kf)^{1/2}} < \infty. \tag{5.2}$$

Indeed, let $\varepsilon = 1/2$ and choose the corresponding δ. Set

$$B = \{r \in \mathbb{R} : |r| \leq \Phi^{-1}(1 - \frac{\delta}{2})(f, Kf)^{1/2}\}.$$

Then $\mu_f(\mathbb{R} \backslash B) = \delta$, whence $\nu_f(\mathbb{R} \backslash B) \leq 1/2$ and $\nu_f(B) \geq 1/2$.
On the other hand, by letting

$$B' = \{r \in \mathbb{R} : |r - (f, h)| \leq 2(f, Kf)^{1/2})\},$$

we have $\nu_f(B') > 1/2$. Therefore, B and B' have a non-empty intersection, i.e.

$$|(f, h)| \leq \left(2 + \Phi^{-1}(1 - \frac{\delta}{2})\right)(f, Kf)^{1/2}$$

which confirms (5.2). Taking into account $(f, Kf) = |I^* f|^2$, from (5.2) we derive that

$$\sup_{f \in \mathscr{X}^*, f \neq 0} \frac{|(f, h)|}{|I^* f|} < \infty.$$

This means that a linear functional $\mathscr{A}$, initially given on a dense subspace $I^* \mathscr{X}^* \subset \mathscr{X}_P^*$ by $\mathscr{A}(I^* f) = (f, h)$, admits an extension to a linear continuous functional on $\mathscr{X}_P^*$. F. Riesz theorem stating a general form for linear continuous functionals on a Hilbert space yields the existence of an element $z \in \mathscr{X}_P^*$ such that

$$(f, h) = \mathscr{A}(I^* f) = (I^* f, z) = (f, Iz), \quad f \in \mathscr{X}^*.$$

It follows that $h = Iz$.   □

The following estimate due to Christer Borell is a nice illustration of virtues of Cameron–Martin theorem.

Proposition 5.1 (Borell inequality for shifted sets [27]) *Let P be a centered Gaussian measure on a linear space $\mathscr{X}$. Let $A \subset \mathscr{X}$ be a symmetric set, i.e. $A = -A$. Then for any $h \in H_P$ it is true that*

$$P(A + h) \geq P(A) \exp \left\{ -\frac{|h|_{H_P}^2}{2} \right\}. \tag{5.3}$$

This estimate is remarkable for its simplicity and even more for its generality.

Proof (of Proposition 5.1). By the symmetry of P and A we have $P(A + h) = P(-A + h) = P(A - h)$. Therefore,

$$
\begin{aligned}
P(A + h) &= \frac{1}{2}(P(A + h) + P(A - h)) \\
&= \frac{1}{2}(P_{-h}(A) + P_h(A)) \\
&= \exp\left(-\frac{|h|^2_{H_P}}{2}\right) \int_A \frac{e^{-z(x)} + e^{z(x)}}{2} P(dx) \\
&\geq \exp\left(-\frac{|h|^2_{H_P}}{2}\right) P(A).
\end{aligned}
$$

$\square$

Remark 5.1 Later on, we will show that for symmetric *convex* sets a converse inequality $P(A) \geq P(A + h)$ is true.

6 Isoperimetric Inequality

6.1 Euclidean Space

Isoperimetric problems are familiar to us from high school courses of physics or geometry in the following form: "among the bodies of given volume, find those of smallest surface area" or "among the bodies of given surface area find those of maximal volume". For Euclidean volume, balls solve both problems. The corresponding isoperimetric inequality can be stated as follows: if a $A \subset \mathbb{R}^n$ is "smooth enough", and B is a ball in $\mathbb{R}^n$, then

$$
\lambda^n(A) = \lambda^n(B) \curvearrowright |\partial A| \geq |\partial B|,
$$

where ∂ on the right-hand side stands for the boundary of a set, and $|\cdot|$ denotes the surface measure. This assertion is not convenient for infinite-dimensional extensions because the notion of surface measure is not robust enough. Therefore, we restate the isoperimetric inequality in a different form. Let B_r be the closed ball of radius r centered at the origin. Define *r-enlargement* of a set A by

$$
A^r := A + B_r.
$$

By replacing the surface measure with the volume of a band of a small width r around a set, we obtain another form of isoperimetric inequality

$$
\lambda^n(A) = \lambda^n(B) \curvearrowright \lambda^n(A^r \setminus A) \geq \lambda^n(B^r \setminus B),
$$

which is equivalent to

$$\lambda^n(A) = \lambda^n(B) \curvearrowright \lambda^n(A^r) \geq \lambda^n(B^r). \tag{6.1}$$

Remarkably, this is true *for all r > 0*, and not only for the small ones.

Remark 6.1 The isoperimetric inequality in Euclidean space is essentially known from antic times. In modern mathematics, requiring more rigor, its history goes back to Steiner [158]. For detailed treatment, see [33].

The geometric statement (6.1) can be easily translated in algebraic language where the ball is not mentioned at all. By using a formula for the volume of Euclidean ball

$$\lambda^n(B_R) = c_n R^n, \quad c_n = \frac{\pi^{n/2}}{\Gamma(\frac{n}{2} + 1)},$$

we see that the radius of a ball B having the same volume as A, is equal to $\left(\frac{\lambda^n(A)}{c_n}\right)^{1/n}$, while the radius of a ball B^r equals to $\left(\frac{\lambda^n(A)}{c_n}\right)^{1/n} + r$. It follows that

$$\lambda^n(A^r) \geq c_n \left(\left(\frac{\lambda^n(A)}{c_n}\right)^{1/n} + r \right)^n,$$

which is equivalent to

$$\left(\frac{\lambda^n(A^r)}{c_n}\right)^{1/n} \geq \left(\frac{\lambda^n(A)}{c_n}\right)^{1/n} + r.$$

The function $\varphi(v) := \left(\frac{v}{c_n}\right)^{1/n}$ is called *isoperimetric function* of the space $(\mathbb{R}^n, \lambda^n)$.

6.2 Euclidean Sphere

Euclidean sphere $\mathbb{S}^n$ is the next important example of a space with isoperimetric property. It is well known that a natural (geodesic) distance $\rho_n(\cdot, \cdot)$ and a unique (up to a constant) finite rotation-invariant measure σ^n are defined on the sphere. Since the sphere is not a vector space, the enlargement of a set $A \subset \mathbb{S}^n$ should be understood as

$$A^r := \{ x \in \mathbb{S}^n : \inf_{y \in A} \rho(x, y) \leq r \}$$

(in Euclidean space this definition is equivalent to the previous one). Isoperimetric inequality on the sphere, due to P. Lévy, asserts (see [33]) that, analogously to Euclidean space case, a ball B provides a solution to isoperimetric problem, i.e.

$$\sigma^n(A) = \sigma^n(B) \curvearrowright \sigma^n(A^r) \geq \sigma^n(B^r).$$

Since the enlargement of a ball on Euclidean sphere is again a ball, we can represent isoperimetric inequality on the sphere in algebraic form $\tilde{\varphi}(\sigma^n(A^r)) \geq \tilde{\varphi}(\sigma^n(A)) + r$ with appropriate isoperimetric function φ.

It is important to notice that the balls in $\mathbb{S}^n$ can be described as "hats", i.e. intersections of $\mathbb{S}^n$ with Euclidean half-spaces. In the following we will consider a limiting passage from $(\mathbb{S}^n, \sigma^n)$ to Gaussian distributions. It will be, therefore, not too surprising that a half-space turns out to be a solution to isoperimetric problem in Gaussian case.

6.3 Poincaré Construction

We will demonstrate a passage from uniform distributions on appropriately chosen spheres to the standard Gaussian distribution in $\mathbb{R}^n$ (this construction is attributed to a great French mathematician H. Poincaré). In particular, by using this construction, one can transfer isoperimetric inequality to the Gaussian case.

Let $m \geq n$. Denote by $\pi_{m,n}$ a natural projection $\mathbb{R}^m \mapsto \mathbb{R}^n$, i.e.

$$\pi_{m,n}(x_1, \ldots, x_m) = (x_1, \ldots, x_n).$$

Let σ^m be the unit rotation-invariant measure on the sphere $\mathbb{S}^m$ in $\mathbb{R}^m$ of radius $\sqrt{m}$, centered at the origin. Define a projection $\nu_{m,n} = \sigma^m \pi_{m,n}^{-1}$ of the measure σ^m, as a measure on $\mathbb{R}^n$ defined by

$$\nu_{m,n}(A) = \sigma^m\left(\pi_{m,n}^{-1}(A)\right).$$

Proposition 6.1 (Poincaré lemma) *Measures $\nu_{m,n}$ weakly converge in $\mathbb{R}^n$ to the standard Gaussian distribution $N(0, E_n)$, as $m \to \infty$.*

Proof. Let $X_1, \ldots, X_n, \ldots, X_m$ be independent standard normal random variables and $\mathbf{X^m} = (X_1, \ldots, X_m) \in \mathbb{R}^m$. Then σ^m is a distribution of the random vector $\left(\frac{X_j\sqrt{m}}{\|\mathbf{X^m}\|}\right)_{1 \leq j \leq m}$ in $\mathbb{R}^m$, while $\nu_{m,n}$ is a distribution of vector $\left(\frac{X_j\sqrt{m}}{\|\mathbf{X^m}\|}\right)_{1 \leq j \leq n}$ in $\mathbb{R}^n$. It remains to notice that $\left(X_j\right)_{1 \leq j \leq n}$ has a distribution $N(0, E_n)$, and that correction factor converges to one by the law of large numbers

$$\frac{\sqrt{m}}{\|\mathbf{X^m}\|} = \sqrt{\frac{m}{X_1^2 + \cdots + X_m^2}} \Rightarrow 1.$$

$\square$

6.4 Euclidean Space with Gaussian Measure

Consider now an isoperimetric problem in the space $\mathbb{R}^n$, equipped with the standard Gaussian measure $P = N(0, E_n)$. The notion of r-enlargement of a set remains the same as in the case of Lebesgue measure. It was shown independently and almost simultaneously by Borell [24] and by Sudakov and Tsirelson [165], that the solution of isoperimetric problem is given by a *half-space* Π, i.e.

$$P(A) = P(\Pi) \curvearrowright P(A^r) \geq P(\Pi^r). \tag{6.2}$$

Let us give an algebraic representation of this inequality. Let us choose Π as

$$\Pi = \left\{ x \in \mathbb{R}^n : x_1 \leq \Phi^{-1}(P(A)) \right\},$$

where $\Phi^{-1}(\cdot)$ is the inverse function to the distribution function of the standard normal law $\Phi(\cdot)$. Clearly,

$$\Pi^r = \left\{ x \in \mathbb{R}^n : x_1 \leq \Phi^{-1}(P(A)) + r \right\},$$

whence $P(\Pi^r) = \Phi\left(\Phi^{-1}(P(A)) + r \right)$, and we may rewrite (6.2) as

$$\Phi^{-1}(P(A^r)) \geq \Phi^{-1}(P(A)) + r. \tag{6.3}$$

We see that $\Phi^{-1}(\cdot)$ is an isoperimetric function for the standard normal distribution.

Proof (of isoperimetric inequality (6.3)). We use Poincaré construction and the corresponding notation. Let $A \subset \mathbb{R}^n$ and let $\Pi_{\tilde{q}} = \{x \in \mathbb{R}^n : x_1 \leq \tilde{q}\}$ be a family of half-spaces. By choosing parameter $q = \Phi^{-1}(P(A))$, we obtain $P(\Pi_q) = P(A)$. Consider a set $A_m = \pi_{m,n}^{-1}(A)$ on the sphere $\mathbb{S}^m$ and the "hats" (balls)

$$B_{m,\tilde{q}} = \{x \in \mathbb{S}^m : x_1 \leq \tilde{q}\} = \pi_{m,n}^{-1}(\Pi_{\tilde{q}}).$$

In particular, let

$$B_m := \{x \in \mathbb{S}^m : x_1 \leq q_m\},$$

where the parameter q_m is chosen so that $\sigma^m(B_m) = \sigma^m(A_m)$.

Poincaré Lemma yields

$$\lim_{m \to \infty} \sigma^m \left(B_{m,\tilde{q}} \right) = \Phi(\tilde{q}) < P(A), \quad \forall \tilde{q} < q.$$

By comparing this with relation

$$\lim_{m\to\infty}\sigma^m\left(B_m\right)=\lim_{m\to\infty}\sigma^m\left(A_m\right)=\lim_{m\to\infty}\sigma^m\left(\pi_{m,n}^{-1}(A)\right)=P(A),$$

we see that for any $\tilde q<q$ it is true that

$$B_m\supset B_{m,\tilde q}\tag{6.4}$$

for all m large enough. Our next argument is a distance contraction under projection. Let $\rho_m(\cdot,\cdot)$ denote the geodesic distance on the sphere $\mathbb{S}^m$. We have

$$\rho_m(x,y)\geq||x-y||\geq||\pi_{m,n}x-\pi_{m,n}y||,\quad\forall x,y\in\mathbb{S}^m.$$

It follows that there is a relation between the enlargements of sets,

$$(A_m)^r\subset\pi_{m,n}^{-1}\left(A^r\right).\tag{6.5}$$

By using consequently Poincaré Lemma, (6.5), Lévy's isoperimetric inequality, and (6.4), we obtain

$$P(A^r)=\lim_{m\to\infty}\sigma^m\pi_{m,n}^{-1}(A^r)\geq\liminf_{m\to\infty}\sigma^m\left((A_m)^r\right)$$

$$\geq\liminf_{m\to\infty}\sigma^m\left((B_m)^r\right)\geq\liminf_{m\to\infty}\sigma^m\left((B_{m,\tilde q})^r\right).$$

We will now use the following property, essentially showing that a sphere of large radius is almost plain: for any $r,\varepsilon>0$ and $\tilde q\in\mathbb{R}$ we have

$$(B_{m,\tilde q})^r\supset B_{m,\tilde q+r-\varepsilon},$$

whenever m is large enough. It follows that

$$\liminf_{m\to\infty}\sigma^m\left((B_{m,\tilde q})^r\right)\geq\liminf_{m\to\infty}\sigma^m\left(B_{m,\tilde q+r-\varepsilon}\right)=\liminf_{m\to\infty}\sigma^m\pi_{m,n}^{-1}(\Pi_{\tilde q+r-\varepsilon})$$

$$=P\left(\Pi_{\tilde q+r-\varepsilon}\right)=\Phi(\tilde q+r-\varepsilon).$$

Therefore,

$$P(A^r)\geq\Phi(\tilde q+r-\varepsilon).$$

By letting $\varepsilon\to0$, $\tilde q\to q$, we obtain (6.2). $\square$

Let now $P=N(0,K)$ be arbitrary Gaussian measure in $\mathbb{R}^n$. Assume that operator K is non-degenerate. Let us choose a strategy of an "intelligent conformism": instead of proving something new, we modify the notions in a way that enables to take advantage of the result obtained for the standard Gaussian measure. Recall that one can construct an $N(0,K)$-distributed vector as $Y=LX$, where X is a random vector having the standard Gaussian distribution and $L=K^{1/2}$. Let $D=\{x\in R^n:||L^{-1}x||\leq1\}$ be the L-image of the unit Euclidean ball. Recall that D coincides with the unit ball (dispersion ellipsoid) of the kernel related to $N(0,K)$ (see Example 4.3). Let us now redefine the notion of r-enlargement in agreement with our situation. Namely, let $A^r:=A+rD$. Let us prove that with this definition inequality (6.3) remains true without any change. Indeed,

$$\Phi^{-1}(P(A^r)) = \Phi^{-1}(\mathbb{P}(Y \in A^r)) = \Phi^{-1}(\mathbb{P}(LX \in A^r))$$

$$= \Phi^{-1}(\mathbb{P}(X \in L^{-1}(A^r))) = \Phi^{-1}(\mathbb{P}(X \in L^{-1}(A) + L^{-1}(rD)))$$

$$= \Phi^{-1}(\mathbb{P}(X \in L^{-1}(A) + B_r)) = \Phi^{-1}(\mathbb{P}(X \in (L^{-1}(A))^r))$$

$$\geq \Phi^{-1}(\mathbb{P}(X \in (L^{-1}(A)))) + r = \Phi^{-1}(\mathbb{P}(Y = LX \in A)) + r$$

$$= \Phi^{-1}(P(A)) + r.$$

6.5 General Linear Space with Gaussian Measure

Let now $X \in \mathscr{X}$ be an arbitrary centered Gaussian random vector in a linear space and let P, resp. H_P, be the corresponding Gaussian distribution and the kernel. Let $D = \{h \in H : \|h\|_{H_P} \leq 1\}$ be the unit ball of the kernel. We define r-enlargement of a set $A \subset \mathscr{X}$ by $A^r := A + rD$. Even in this general situation we are able to reproduce a result that we obtained earlier for $\mathscr{X} = \mathbb{R}^n$.

Theorem 6.1 (Gaussian isoperimetric inequality) *For any measurable $A \subset \mathscr{X}$ it is true that*

$$\Phi^{-1}(P_*(A^r)) \geq \Phi^{-1}(P(A)) + r, \tag{6.6}$$

where interior measure $P_*(C)$ is defined by the formula

$$P_*(C) := \sup_{B \text{ compact}, B \subset C} P(B). \tag{6.7}$$

for any $C \subset \mathscr{X}$. The equality in (6.6) is attained on half-spaces.

Remark 6.2 One can get rid of interior measure by writing the same assertion as

$$B \cap A^r = \emptyset \curvearrowright P(B) \leq \widehat{\Phi}\left(\Phi^{-1}(P(A)) + r\right), \tag{6.8}$$

for any measurable B, where $\widehat{\Phi}(\cdot) = 1 - \Phi(\cdot)$ stands for the tail of standard normal law.

6.6 Concentration Principle

Concentration Principle is a remarkable corollary of isoperimetric inequality. It asserts that the distribution concentration for Lipschitz functionals of a Gaussian vector is at least as strong as that of univariate normal law.

A functional $f : \mathscr{X} \mapsto \mathbb{R}$ is called H_P-*Lipschitz* with a constant σ, if

$$|f(x+h) - f(x)| \leq \sigma |h|_{H_P}, \quad \forall x \in \mathscr{X}, h \in H_P.$$

In this case we write $f \in Lip_{H_P}(\sigma)$.

Example 6.1 (*Supremum as a Lipschitz functional*) Let $\mathscr{X} = \mathbb{C}(T)$ and let $X = (X(t))_{t \in T} \in \mathscr{X}$ be a centered Gaussian process. Consider a functional

$$f(x) = \sup_{t \in T} x(t).$$

Then for any $x \in \mathscr{X}, h = Iz \in H_P$, it is true that

$$|f(x+h) - f(x)| = \left| \sup_{t \in T} x(t) - \sup_{t \in T}(x(t) + h(t)) \right| \leq \sup_{t \in T} |h(t)|$$
$$= \sup_{t \in T} |(\delta_t, Iz)| = \sup_{t \in T} |(I^*\delta_t, z)_{\mathscr{X}_P^*}| \leq \sup_{t \in T} ||I^*\delta_t||_{\mathscr{X}_P^*} ||z||_{\mathscr{X}_P^*}$$
$$= \sup_{t \in T} ||I^*\delta_t||_{\mathscr{X}_P^*} |h|_{H_P} = \sigma |h|_{H_P},$$

where

$$\sigma^2 = \sup_{t \in T} ||I^*\delta_t||^2_{\mathscr{X}_P^*} = \sup_{t \in T} \mathbb{E}X(t)^2.$$

Exercise 6.1 Let X be a centered Gaussian random vector in a separable Banach space $(\mathscr{X}, ||\cdot||)$, let $K : \mathscr{X}^* \mapsto \mathscr{X}$ be the covariance operator of X. Prove that the functional $f(x) = ||x||$ belongs to the class $Lip_{H_P}\left(\sqrt{||K||}\right)$.

Recall that a number m is called a *median* of the distribution of a functional f, if

$$\mathbb{P}(f(X) \leq m) \geq \frac{1}{2} \text{ and } \mathbb{P}(f(X) \geq m) \geq \frac{1}{2}.$$

A median need not be unique but this is irrelevant to the following bound.

Theorem 6.2 (Concentration Principle) *If* $f \in Lip_{H_P}(\sigma)$ *and* m *is a median of* f, *then for any* $r > 0$ *it is true that*

$$\mathbb{P}(f(X) \geq m + r) \leq \hat{\Phi}\left(\frac{r}{\sigma}\right), \tag{6.9}$$

$$\mathbb{P}(f(X) \leq m - r) \leq \hat{\Phi}\left(\frac{r}{\sigma}\right). \tag{6.10}$$

Equalities are attained if f *is a linear functional.*

Proof. Let $A = \{x \in \mathscr{X} : f(x) \leq m\}$. We plug the relation $P(A) \geq \frac{1}{2}$ in isoperimetric inequality (6.6) and obtain

$$\Phi^{-1}(P_*(A^{\frac{r}{\sigma}})) \geq \Phi^{-1}(P(A)) + \frac{r}{\sigma} \geq \frac{r}{\sigma}. \tag{6.11}$$

On the other hand, for any $y \in A^{\frac{r}{\sigma}}$ there is a representation $y = x + h$ where $x \in A$, $|h|_{H_P} \leq \frac{r}{\sigma}$. Therefore,

$$f(y) \leq f(x) + \sigma |h|_{H_P} \leq m + r.$$

This can be represented as

$$\{x \in \mathscr{X} : f(x) > m + r\} \cap A^{\frac{r}{\sigma}} = \emptyset,$$

whence by (6.11),

$$\mathbb{P}(f(X) > m + r) \leq 1 - P_* \left(A^{\frac{r}{\sigma}} \right) \leq 1 - \Phi \left(\frac{r}{\sigma} \right) = \widehat{\Phi} \left(\frac{r}{\sigma} \right).$$

By applying (6.9) to the functional $-f$, we obtain (6.10). □

Theorem 6.2 shows that functional distributions can be controlled through two parameters, m and σ. Once the concentration bounds are obtained, it is easy to evaluate the moment characteristics of a functional. Here are two simple examples.

Corollary 6.1 *If* $f \in Lip_{H_P}(\sigma)$, *then*

$$\mathbb{E} \exp\{\alpha |f(X)|^2\} < \infty, \quad 0 \leq \alpha < \frac{1}{2\sigma^2}.$$

Corollary 6.2 *If* $f \in Lip_{H_P}(\sigma)$, *then*

$$\mathbb{V}ar f(X) \leq \sigma^2.$$

Indeed, by a standard property of the variance,

$$\mathbb{V}ar f(X) \leq \mathbb{E}(f(X) - m)^2 = \int_0^\infty \mathbb{P} \left(|f(X) - m| \geq r^{1/2} \right) dr$$
$$\leq 2 \int_0^\infty \widehat{\Phi} \left(\frac{r^{1/2}}{\sigma} \right) dr = \sigma^2.$$

There are several alternative approaches to isoperimetric inequalities. One of them, based on symmetrization transformations, is due to Ehrhard [59], see also a detailed exposition in [117]. An approach related to functional inequalities is due to Bobkov [20]. Finally, for a more elementary approach, see another Bobkov's work [21].

There exists an extensive literature about isoperimetric inequalities and related concentration inequalities for *non-Gaussian* measures, see [105, 106, 171, 172] and references therein.

7 Measure Concavity and Other Inequalities

7.1 Measure Concavity

A function $\varphi : \mathscr{X} \mapsto \mathbb{R}$ defined on a linear space $\mathscr{X}$, is called *concave*, if

$$\varphi(\alpha x + (1 - \alpha)y) \geq \alpha\varphi(x) + (1 - \alpha)\varphi(y)$$

for all $x, y \in \mathscr{X}$ and $\alpha \in [0, 1]$. A similar notion can be introduced for measures. The property

$$\mu(\alpha A + (1 - \alpha)B) \geq \alpha\mu(A) + (1 - \alpha)\mu(B),$$

would be the most natural extension. Unfortunately, the interesting measures μ having this property do not exist. It is, however, possible to introduce a notion of concavity depending on some function Q:

$$\mu(\alpha A + (1 - \alpha)B) \geq Q(\mu(A), \mu(B), \alpha)$$

for all measurable $A, B \subset \mathscr{X}$ and all $\alpha \in [0, 1]$. For example, the famous Brunn–Minkowski inequality [33, 75] asserts that for Lebesgue measure in $\mathbb{R}^n$ it is true that

$$\left[\lambda^n(\alpha A + (1 - \alpha)B)\right]^{1/n} \geq \alpha\left[\lambda^n(A)\right]^{1/n} + (1 - \alpha)\left[\lambda^n(B)\right]^{1/n}. \tag{7.1}$$

Let us consider two more versions of concavity—logarithmic concavity and Ehrhard concavity.

A measure μ is called *logarithmically concave*, if

$$\ln \mu(\alpha A + (1 - \alpha)B) \geq \alpha \ln \mu(A) + (1 - \alpha) \ln \mu(B) \tag{7.2}$$

or, equivalently,

$$\mu(\alpha A + (1 - \alpha)B) \geq \mu(A)^\alpha \mu(B)^{1-\alpha}.$$

By combining Brunn–Minkowski inequality (7.1) with concavity of logarithmic function we see that Lebesgue measure in $\mathbb{R}^n$ satisfies (7.2). The advantage of logarithmic concavity (7.2) when compared with (7.1) is the dimension invariance of (7.2), which enables to extend this property to infinite-dimensional case.

Borell [25] proved that any measure in $\mathbb{R}^n$ having a density $p(x) = e^{\varphi(x)}$, is logarithmically convex whenever $\varphi(\cdot)$ is a concave function. In particular, for a Gaussian measure $N(a, K)$ with non-degenerated operator K

$$\varphi(x) = \ln\left((2\pi)^{-n/2}(det K)^{-1/2}\right) - \frac{1}{2}\left(K^{-1}(x - a), (x - a)\right)$$

is a concave quadratic function. Easy passages to the limit show that any Gaussian measure in a linear space is logarithmically concave.

Since the Gaussianity is not something special from the point of view of the definition of logarithmical convexity, it is not surprising that *equality* in (7.2) for Gaussian measure is not attained on any non-trivial sets.

A measure μ is called *Ehrhard concave*, if

$$\Phi^{-1}\left(\mu_*(\alpha A + (1 - \alpha)B)\right) \geq \alpha \Phi^{-1}\left(\mu(A)\right) + (1 - \alpha)\Phi^{-1}\left(\mu(B)\right). \quad (7.3)$$

Here $\Phi^{-1}(\cdot)$, as usual, is the function inverse to the distribution function of standard normal law and μ_* on the left-hand side is internal measure defined in (6.7). Any Gaussian measure is Ehrhard concave.[3] The equality in (7.3) is attained on half-spaces having parallel boundaries.

Any kind of concavity yields the following useful result.

Corollary 7.1 (Anderson inequality [3]). *Let P be a centered Gaussian measure in a linear space $\mathscr{X}$ and let A be a convex symmetric subset of $\mathscr{X}$. Then for any $h \in \mathscr{X}$ it is true that*

$$P(A + h) \leq P(A).$$

Proof. Let $A' = A + h$ and $B' = A - h$. Since A is symmetric, we have

$$B' = A - h = -A - h = -(A + h) = -A'.$$

By the symmetry of P it follows $P(A') = P(B')$. We apply (7.2) to the sets (A') and (B') and obtain

$$\ln P\left(\frac{A' + B'}{2}\right) \geq \frac{1}{2}\ln P(A') + \frac{1}{2}\ln P(B') = \ln P(A').$$

Finally, convexity of A yields

$$\frac{A' + B'}{2} = \frac{A + h + A - h}{2} = \frac{A + A}{2} = A.$$

Therefore, $P(A) \geq P(A') = P(A + h)$.   □

Example 7.1 (*Sample paths running near a given curve*) Let $X(t), t \in T$, be a centered Gaussian process and $\varepsilon : T \mapsto \mathbb{R}_+$. Consider a symmetric convex set of functions

$$A = \{x : T \mapsto \mathbb{R}, |x(t)| \leq \varepsilon(t), \forall t \in T\}.$$

---

[3] This is a highly non-trivial result. Ehrhard [59] proved (7.3) in 1983 for convex sets. In 1996 Latała [102] proved it for the case when only one of the sets is assumed to be convex. Finally, Borell [28] investigated the general case in 2003. For further improvements see [13, 29, 76].

According to Anderson inequality, for any function $h : T \mapsto \mathbb{R}$ it is true that

$$\mathbb{P}(X \in A) = \mathbb{P}\{|X(t)| \le \varepsilon(t), \forall t \in T\} \ge \mathbb{P}\{|X(t) - h(t)| \le \varepsilon(t), \forall t \in T\}.$$

Therefore, a sample path of Gaussian process runs through a strip near its expectation with a larger probability than through a strip of the same shape near other given curve.

Concavity can also be applied to the investigation of the distributions of convex functionals of a Gaussian vector X. Let $\varphi : \mathscr{X} \mapsto \mathbb{R}$ be a convex function, i.e.

$$\varphi(\alpha x + (1 - \alpha)y) \le \alpha\varphi(x) + (1 - \alpha)\varphi(y)$$

for all $x, y \in \mathscr{X}$ and $\alpha \in [0, 1]$. Let $F(r) = \mathbb{P}\{\varphi(X) \le r\}$ denote the distribution function of the functional $\varphi(X)$.

Proposition 7.1 *The function*

$$r \mapsto \Phi^{-1}(F(r))$$

is concave.

Proof. For any $r_1, r_2 \in \mathbb{R}$ we let

$$A = \{x \in \mathscr{X} : \varphi(x) \le r_1\}, \quad B = \{x \in \mathscr{X} : \varphi(x) \le r_2\}.$$

Then for any $\alpha \in [0, 1]$ it is true that

$$\alpha A + (1 - \alpha)B \subset \{x \in \mathscr{X} : \varphi(x) \le \alpha r_1 + (1 - \alpha)r_2\}.$$

Ehrhard inequality (7.3) yields

$$\begin{aligned}
\Phi^{-1}(F(\alpha r_1 + (1 - \alpha)r_2)) &= \Phi^{-1}(\mathbb{P}\{\varphi(X) \le \alpha r_1 + (1 - \alpha)r_2\}) \\
&\ge \Phi^{-1}(\mathbb{P}\{X \in \alpha A + (1 - \alpha)B\}) \\
&\ge \alpha\Phi^{-1}(\mathbb{P}\{X \in A\}) + (1 - \alpha)\Phi^{-1}(\mathbb{P}\{X \in B\}) \\
&= \alpha\Phi^{-1}(F(r_1)) + (1 - \alpha)\Phi^{-1}(F(r_2)).
\end{aligned}$$

$\square$

It follows from this proposition that the distribution of random variable $\varphi(X)$ has approximately the same degree of regularity as a convex function of real variable. Namely, if Q denotes the distribution of $\varphi(X)$, than Q is absolutely continuous with respect to Lebesgue measure, except for an eventual atom at the starting point $r_* = \inf\{r : F(r) > 0\}$. On the interval (r_*, ∞) measure Q has a density which is continuous everywhere except for a countable number of points where it has downward jumps.

Corollary 7.2 *Let m be a median for the distribution of $\varphi(X)$. Then*

$$m \leq \mathbb{E}\varphi(X). \tag{7.4}$$

Proof. For the sake of simplicity, let us assume that the distribution of $\varphi(X)$ has no atom. Then $F(m) = \frac{1}{2}$ and the variable $F(\varphi(X))$ is uniformly distributed on the interval [0,1]. By the definition, $\Phi^{-1}(F(m)) = \Phi^{-1}\left(\frac{1}{2}\right) = 0$. On the other hand, the concavity of $\Phi^{-1}(F(\cdot))$ and Jensen inequality yield

$$\Phi^{-1}(F(\mathbb{E}\varphi(X))) \geq \mathbb{E}\Phi^{-1}(F(\varphi(X))) = \int_0^1 \Phi^{-1}(p)\mathrm{d}p = 0.$$

Corollary follows now from monotonicity of the function $\Phi^{-1}(F(\cdot))$. $\qquad\square$

7.2 Dilations

Many Gaussian inequalities compare the measures of sets A and rA, where A is a symmetric convex set. So called *S-property* is one of the most famous results. It was stated as *S-conjecture* by Kwapień and Sawa [101] and later proved by Latała and Oleszkiewicz [104]. *S*-property demonstrates extremal features of a strip

$$S_{r,f} := \{x \in \mathscr{X} : |(f,x)| \leq r\}, \quad r > 0, f \in \mathscr{X}^*.$$

Let P be the distribution of a centered Gaussian vector X. Without loss of generality we may assume that $\mathbb{E}(f, X)^2 = 1$. Then

$$P\left(S_{r,f}\right) = 2\Phi(r) - 1 := G(r).$$

S-property asserts that among all symmetric convex sets the strip demonstrates the slowest mass loss under compression and the slowest mass increase under dilation. Analytically this means that

$$P(A) = G(\rho) \curvearrowright P(rA) \geq G(r\rho), \quad r \geq 1;$$
$$P(A) = G(\rho) \curvearrowright P(rA) \leq G(r\rho), \quad r \leq 1.$$

Such properties are widely used in geometric applications of probability theory but they have no decent applications in the theory of random processes because the bound $G(r\rho)$ decreases polynomially, while $P(rA)$ decreases exponentially, as $r \to 0$, for typical sets related to random processes.

Let us finally mention a more special notion of *B-concavity*, introduced by W. Banaszczyk and finally established by Cordero-Erausquin et al. [40]. When compared to the types of concavity that we considered earlier, it is related to a narrow class of sets and can be stated as follows.

Theorem 7.1 *Let P be a centered Gaussian distribution in $\mathscr{X}$ and let A be a centrally symmetric convex set in $\mathscr{X}$. Then the function*

$$t \mapsto \ln P(e^t A)$$

is concave on $(0, \infty)$.

This theorem yields, for example, that

$$P\left(e^{\frac{t_1+t_2}{2}} A\right) \geq P\left(e^{t_1} A\right)^{1/2} P\left(e^{t_2} A\right)^{1/2}.$$

By plugging $c_1 = e^{t_1}$, $c_2 = e^{t_2}$ in, we obtain

$$P\left(\sqrt{c_1 c_2} A\right) \geq P\left(c_1 A\right)^{1/2} P\left(c_2 A\right)^{1/2},$$

while the usual logarithmic concavity (7.2) yields a weaker inequality

$$P\left(\frac{c_1 + c_2}{2} A\right) = P\left(\frac{1}{2}(c_1 A) + \frac{1}{2}(c_2 A)\right) \geq P\left(c_1 A\right)^{1/2} P\left(c_2 A\right)^{1/2}.$$

7.3 Correlation Conjecture

If X_1 and X_2 are two independent random vectors in $\mathscr{X}$, then for any measurable sets $A_1, A_2 \subset \mathscr{X}$ it is true that

$$\mathbb{P}(X_1 \in A_1, X_2 \in A_2) = \mathbb{P}(X_1 \in A_1)\,\mathbb{P}(X_2 \in A_2).$$

When handling dependent *Gaussian* vectors, we can still affirm that for the sets A_1, A_2 from a certain class it is true that

$$\mathbb{P}(X_1 \in A_1, X_2 \in A_2) \geq \mathbb{P}(X_1 \in A_1)\,\mathbb{P}(X_2 \in A_2). \qquad (7.5)$$

The famous *correlation conjecture*, arguably the most attractive open problem in the theory of Gaussian processes, asserts that (7.5) holds whenever X_1, X_2 are centered Gaussian vectors and A_1, A_2 are symmetric convex sets. We refer to [154] for the history and various equivalent formulations.

Correlation conjecture also admits a slightly different representation. Let X be a centered Gaussian vector and let $\mathscr{A}_1, \mathscr{A}_2$ be symmetric convex sets. Then

$$\mathbb{P}(X \in \mathscr{A}_1 \cap \mathscr{A}_2) \geq \mathbb{P}(X \in \mathscr{A}_1)\mathbb{P}(X \in \mathscr{A}_2), \qquad (7.6)$$

which at the first glance may look as a kind of paradoxical independence of X of itself. The connection between two kinds of inequalities becomes clear if we let $X = (X_1, X_2) \in \mathscr{X} \times \mathscr{X}$ and define symmetric convex sets by

$$\mathscr{A}_1 = \{(x_1, x_2) \in \mathscr{X} \times \mathscr{X} : x_1 \in A_1\}, \quad \mathscr{A}_2 = \{(x_1, x_2) \in \mathscr{X} \times \mathscr{X} : x_2 \in A_2\}.$$

Then (7.6) becomes (7.5). On the other hand, by plugging $X = X_1 = X_2$ in (7.5), we obtain (7.6).

Correlation conjecture (7.6) is proved in some important cases:

- for $\mathscr{X} = \mathbb{R}^2$ (Pitt [149]).
- when one of the sets $\mathscr{A}_1$ or $\mathscr{A}_2$ is a strip, e.g.,

$$\mathscr{A}_2 = \{x \in \mathscr{X} : |(f, x)| \leq r\}, \quad r \geq 0, f \in \mathscr{X}^*$$

(proved independently by Khatri and Šidák [89, 156]).
- when one of the sets $\mathscr{A}_1$ or $\mathscr{A}_2$ is a symmetric ellipsoid (Hargé [80]).
- in some other particular cases ... see [103] for more references.

By iterating the statement for the strips $\mathscr{A}_j = \{x \in \mathscr{X} : |(f_j, x)| \leq r_j\}$, $j = 2, 3, \ldots$, we obtain

$$\mathbb{P}(X \in \mathscr{A}_1 \cap (\mathscr{A}_2 \cap \mathscr{A}_3)) \geq \mathbb{P}(X \in (\mathscr{A}_1 \cap \mathscr{A}_2)) \mathbb{P}(X \in \mathscr{A}_3)$$
$$\geq \mathbb{P}(X \in \mathscr{A}_1) \mathbb{P}(X \in \mathscr{A}_2) \mathbb{P}(X \in \mathscr{A}_3),$$

and, by further iterations, we arrive at

$$\mathbb{P}\left(X \in \mathscr{A}_1 \cap \left(\bigcap_{j=2}^{n} \mathscr{A}_j\right)\right) \geq \mathbb{P}(X \in \mathscr{A}_1) \prod_{j=2}^{n} \mathbb{P}(X \in \mathscr{A}_j).$$

In particular, for $A_1 = \mathscr{X}$ we obtain

$$\mathbb{P}\left(X \in \bigcap_{j=2}^{n} \mathscr{A}_j\right) \geq \prod_{j=2}^{n} \mathbb{P}(X \in \mathscr{A}_j). \tag{7.7}$$

The latter relation is known as *Khatri–Šidák inequality* [89, 156]. It is, of course, also true for intersection of countable number of strips.

Although correlation conjecture is not proved in full generality, the following weaker version is often sufficient for applications. It is due to Schechtman et al. [154], (for special case $1 - \varepsilon = \sqrt{\frac{1}{2}}$) and to Li [111] (for general ε).

Theorem 7.2 (Weak correlation inequality) *For any $\varepsilon \in (0, 1)$ there exists $K_\varepsilon > 0$ such that for all centered Gaussian vectors X and for all symmetric convex sets $\mathscr{A}_1, \mathscr{A}_2$ it is true that*

$$\mathbb{P}(X \in \mathscr{A}_1 \cap \mathscr{A}_2) \geq \mathbb{P}(X \in (1 - \varepsilon)\mathscr{A}_1) \mathbb{P}\left(X \in \frac{\mathscr{A}_2}{K_\varepsilon}\right). \tag{7.8}$$

Proof (of Theorem 7.2). Let $\tilde{X}$ be a vector having the same distribution as X and independent of X. It is easy to check that for any $a > 0$ the vectors $X - a\tilde{X}$ and $X + a^{-1}\tilde{X}$ are independent. To justify this claim, it is sufficient to check that the values of arbitrary linear functionals of these Gaussian vectors are non-correlated, and we have indeed

$$\mathbb{E}(f, X - a\tilde{X})(g, X + a^{-1}\tilde{X})$$
$$= \mathbb{E}(f, X)(g, X) + a^{-1}\mathbb{E}(f, X)(g, \tilde{X}) - a\mathbb{E}(f, \tilde{X})(g, X) - \mathbb{E}(f, \tilde{X})(g, \tilde{X})$$
$$= \mathbb{E}(f, X)(g, X) + a^{-1}\mathbb{E}(f, X)\mathbb{E}(g, \tilde{X}) - a\mathbb{E}(f, \tilde{X})\mathbb{E}(g, X) - \mathbb{E}(f, \tilde{X})(g, \tilde{X})$$
$$= 0.$$

Notice that the vectors $X - a\tilde{X}$ and $(1 + a^2)^{1/2}X$, $X + a^{-1}\tilde{X}$ and $(1 + a^{-2})^{1/2}X$ are equidistributed by stability of Gaussian distributions.

Furthermore, by using Anderson inequality for the vector $(X, X) \in \mathscr{X} \times \mathscr{X}$, the sets $\mathscr{A}_1 \times \mathscr{A}_2 \subset \mathscr{X} \times \mathscr{X}$ and for the random shift $(a\tilde{X}, -a^{-1}\tilde{X}) \in \mathscr{X} \times \mathscr{X}$, we obtain

$$\mathbb{P}(X \in \mathscr{A}_1, X \in \mathscr{A}_2) \geq \int_{\mathscr{X}} \mathbb{P}(X \in \mathscr{A}_1 + ah, X \in \mathscr{A}_2 - a^{-1}h)P(dh)$$
$$= \mathbb{P}(X - a\tilde{X} \in \mathscr{A}_1, X + a^{-1}\tilde{X} \in \mathscr{A}_2)$$
$$= \mathbb{P}(X - a\tilde{X} \in \mathscr{A}_1)\mathbb{P}(X + a^{-1}\tilde{X} \in \mathscr{A}_2)$$
$$= \mathbb{P}((1 + a^2)^{1/2}X \in \mathscr{A}_1)\mathbb{P}((1 + a^{-2})^{1/2}X \in \mathscr{A}_2).$$

By plugging $a = \frac{(2\varepsilon - \varepsilon^2)^{1/2}}{1-\varepsilon}$ in this relation, we obtain (7.8) with the constant $K_\varepsilon = (2\varepsilon - \varepsilon^2)^{1/2}$. $\square$

Let us present a typical application of weak correlation to small deviation theory. Let X be a centered Gaussian vector in a normed space $(\mathscr{X}, ||\cdot||)$, admitting a representation $X = X_1 + X_2$, where X_1, X_2 are Gaussian, eventually dependent, vectors such that X_1 is well studied while X_2 is relatively small. By using the triangle inequality (7.8), we obtain

$$\mathbb{P}(||X|| \leq r) = \mathbb{P}(||X_1 + X_2|| \leq r)$$
$$\geq \mathbb{P}(||X_1|| \leq (1 - \varepsilon)r, ||X_2|| \leq \varepsilon r)$$
$$\geq \mathbb{P}\left(||X_1|| \leq (1 - \varepsilon)^2 r\right)\mathbb{P}\left(||X_2|| \leq \frac{\varepsilon r}{K_\varepsilon}\right).$$

If the vector X_2 is essentially smaller than X_1, the second factor is less important for asymptotic behavior of the probability, as $r \to 0$, than the first one, and the small deviations of the sum reduce to those of one term. The upper bound can easily be obtained from the lower bound. By writing $X_1 = (X_1 + X_2) + (-X_2)$ and replacing r with $(1 - \varepsilon)^{-2}r$, we obtain

$$\mathbb{P}(||X_1|| \leq (1 - \varepsilon)^{-2}r) \geq \mathbb{P}\left(||X_1 + X_2|| \leq r\right)\mathbb{P}\left(||X_2|| \leq \frac{\varepsilon(1 - \varepsilon)^{-2}r}{K_\varepsilon}\right), \quad (7.9)$$

i.e.

$$\mathbb{P}\left(||X_1 + X_2|| \leq r\right) \leq \frac{\mathbb{P}(||X_1|| \leq (1 - \varepsilon)^{-2}r)}{\mathbb{P}\left(||X_2|| \leq \frac{\varepsilon(1-\varepsilon)^{-2}r}{K_\varepsilon}\right)}. \quad (7.10)$$

7.4 Bounds for Shifted Measures

Isoperimetric inequality evaluates how a measure of a set can grow when we pass to the set's ε-enlargement. In this subsection we investigate how can a Gaussian measure of a set change if we *shift* the set. This turns out to be a relatively easy problem.

Theorem 7.3 (*Kuelbs and Li* [97]) *Let X be a centered Gaussian vector in $\mathscr{X}$ with distribution P. Then for any measurable set $A \subset \mathscr{X}$ and any $h \in H_P$ it is true that*

$$\Phi^{-1}(P(A)) - |h|_{H_P} \leq \Phi^{-1}(P(A+h)) \leq \Phi^{-1}(P(A)) + |h|_{H_P}.$$

The equality in both inequalities is attained on the corresponding half-spaces.

Proof (of Theorem 7.3). In order to concentrate attention on essential details we only consider the case $\mathscr{X} = \mathbb{R}^n$, $P = N(0, E)$ and $|h|_{H_P} = 1$. Let us prove the upper bound (the lower bound follows by application of the upper bound to the complement set). Consider the half-space

$$\Pi = \{x \in \mathbb{R}^n : (h, x) \leq r\}, \quad \text{where } r = \Phi^{-1}(P(A)).$$

Then $P(\Pi) = \mathbb{P}((h, X) \leq r) = \Phi(r) = P(A)$ and

$$\Pi + h = \{x \in \mathbb{R}^n : (h, x) \leq r + 1\}.$$

We will check that

$$P(A + h) \leq P(\Pi + h). \tag{7.11}$$

Then we obtain the required relation

$$\Phi^{-1}(P(A+h)) \leq \Phi^{-1}(P(\Pi + h)) = \Phi^{-1}(\mathbb{P}((h, X) \leq r + 1))$$
$$= r + 1 = \Phi^{-1}(P(A)) + 1 = \Phi^{-1}(P(A)) + |h|_{H_P}.$$

By Cameron–Martin formula, we have

$$P(A + h) = P_{-h}(A) = \int_A e^{-\frac{1}{2}-(h,x)} P(dx)$$
$$= \int_{A \cap \Pi} e^{-\frac{1}{2}-(h,x)} P(dx) + \int_{A \setminus \Pi} e^{-\frac{1}{2}-(h,x)} P(dx)$$

and

$$P(\Pi + h) = P_{-h}(\Pi) = \int_\Pi e^{-\frac{1}{2}-(h,x)} P(dx)$$
$$= \int_{A \cap \Pi} e^{-\frac{1}{2}-(h,x)} P(dx) + \int_{\Pi \setminus A} e^{-\frac{1}{2}-(h,x)} P(dx).$$

The first terms of two expressions coincide. By taking into account the relations

$$P(A \setminus \Pi) = P(A) - P(A \cap \Pi) = P(\Pi) - P(A \cap \Pi) = P(\Pi \setminus A)$$

and the definition of Π, we achieve the required inequality for the second terms,

$$\int_{A \setminus \Pi} e^{-(h,x)} P(dx) \le e^{-r} P(A \setminus \Pi) = e^{-r} P(\Pi \setminus A) \le \int_{\Pi \setminus A} e^{-(h,x)} P(dx).$$

Now (7.11) is proved.   □

Exercise 7.1 Repeat the proof for $h \in \mathbb{R}^n$ having arbitrary norm. Next, prove the theorem for arbitrary Gaussian vector in $\mathbb{R}^n$.

For more information about Gaussian and related inequalities, see the surveys [12, 103].

8 Large Deviation Principle

8.1 Cramér–Chernoff Theorem and General Large Deviation Principle

In this section we consider a version of Large Deviation Principle applicable to Gaussian distributions. In full generality, Large Deviation Principle provides an evaluation methodology for probability of large deviation of a random object from its "typical position". It is often used in mathematical statistics for comparison of theoretical distributions and practical data, in statistical mechanics, etc. The following result is a simplest and most representative example of Large Deviation Principle.

Theorem 8.1 (Cramér–Chernoff theorem [38, 41]) *Let* $X_1, \ldots$ *be a sequence of i.i.d. random variables satisfying assumption*

$$\mathbb{E} \exp\{\gamma X_j\} < \infty, \quad |\gamma| < \gamma_0,$$

for some $\gamma_0 > 0$. *Let denote* $S_n = \sum_{j=1}^n X_j$. *Then for any* $r > \mathbb{E} X_1$ *it is true that*

$$\lim_{n \to \infty} \frac{\ln \mathbb{P}\left\{ \frac{S_n}{n} \ge r \right\}}{n} = -I(r),$$

where the function

$$I(r) = \sup_\gamma \{\gamma r - \ln \mathbb{E} e^{\gamma X_1}\}$$

is called deviation function.

The upper bound in Cramér–Chernoff theorem follows from a simple application of exponential Chebyshev inequality

$$\mathbb{P}\left\{\frac{S_n}{n} \ge r\right\} = \mathbb{P}\left\{S_n \ge rn\right\} \le \frac{\mathbb{E}e^{\gamma S_n}}{e^{\gamma rn}}$$

$$= \frac{\left(\mathbb{E}e^{\gamma X_1}\right)^n}{e^{\gamma rn}} = \exp\left\{-n\left(\gamma r - \ln \mathbb{E}e^{\gamma X_1}\right)\right\},$$

with subsequent optimization in γ. However, the lower bound requires more subtle and asymptotical computations.

The multivariate version of Cramér–Chernoff theorem turns out to be more involved.

Theorem 8.2 (Large Deviation Principle in $\mathbb{R}^n$) *Let* $X_1, \cdots \in \mathbb{R}^n$ *be a sequence of i.i.d. random vectors satisfying assumption*

$$\mathbb{E}\exp\{(\gamma, X_j)\} < \infty, \quad \gamma \in \mathbb{R}^n, |\gamma| < \gamma_0,$$

for some $\gamma_0 > 0$. *Let* $S_n = \sum_{j=1}^n X_j$. *Then for any open set* $G \subset \mathbb{R}^n$ *it is true that*

$$\liminf_{n\to\infty} \frac{\ln \mathbb{P}\left\{\frac{S_n}{n} \in G\right\}}{n} \ge -J(G),$$

and for any closed set $F \subset \mathbb{R}^n$ *it is true that*

$$\limsup_{n\to\infty} \frac{\ln \mathbb{P}\left\{\frac{S_n}{n} \in F\right\}}{n} \le -J(F),$$

where $J(A) = \inf_A I(\cdot)$ *and the function* $I : \mathbb{R}^n \mapsto [0, \infty]$ *called deviation function is defined by*

$$I(h) = \sup_{\gamma \in \mathbb{R}^n} \{(\gamma, h) - \ln \mathbb{E}e^{(\gamma, X_1)}\}.$$

Let us apply this theorem to the vectors having the standard Gaussian distribution. Clearly $\ln \mathbb{E}e^{(\gamma, X_1)} = \frac{(\gamma, \gamma)}{2}$, which easily implies $I(h) = \frac{|h|^2}{2}$.

Also notice that in Gaussian case $\frac{S_n}{n} = \frac{X_1}{\sqrt{n}}$ in distribution, thus the result can be stated directly in terms of X_1. Moreover, after the sums are gone, it is natural to replace the integer parameter n by the real one $R = \sqrt{n}$. It follows that for any open $G \subset \mathbb{R}^n$ it is true that

$$\liminf_{R\to\infty} \frac{\ln \mathbb{P}\{X \in RG\}}{R^2} \ge -J(G),$$

and for any closed $F \subset \mathbb{R}^n$ it is true that

$$\limsup_{R \to \infty} \frac{\ln \mathbb{P}\{X \in RF\}}{R^2} \leq -J(F)$$

where $J(A) = \inf_{h \in A} \frac{|h|^2}{2}$.

Finally, recall that for any set $A \subset \mathbb{R}^n$ we have

$$\inf_{Cl(A)} I(\cdot) \leq \inf_A I(\cdot) \leq \inf_{Int(A)} I(\cdot).$$

If

$$\inf_{Cl(A)} I(\cdot) = \inf_A I(\cdot) = \inf_{Int(A)} I(\cdot),$$

then the set A is called *regular*. Under assumptions of Theorem 8.2 for any regular set A we have

$$\lim_{n \to \infty} \frac{\ln \mathbb{P}\left\{\frac{S_n}{n} \in A\right\}}{n} = -J(A),$$

and in Gaussian case

$$\lim_{R \to \infty} \frac{\ln \mathbb{P}\{X \in RA\}}{R^2} = -J(A).$$

Theorem 8.2 motivates the following definition [178] a family of distributions (P_n) in a topological space $\mathscr{X}$ satisfies *Large Deviation Principle* with a rate v_n and deviation function $I : \mathscr{X} \mapsto [0, \infty]$, if for any open set $G \subset \mathscr{X}$ it is true that

$$\liminf_{n \to \infty} \frac{\ln P_n(G)}{v_n} \geq -J(G),$$

and for any closed set $F \subset \mathscr{X}$ it is true that

$$\limsup_{n \to \infty} \frac{\ln P_n(F)}{v_n} \leq -J(F),$$

where $J(A) = \inf_A I(\cdot)$.

The book [47] is an excellent source for learning general large deviation theory.

8.2 Large Deviation Principle for Gaussian Vector

Let now X be a centered Gaussian vector with distribution P in a linear space $\mathscr{X}$. We define the deviation function by a formula analogous to that of finite-dimensional case:

$$I(h) = \begin{cases} \frac{|h|^2_{H_P}}{2}, & h \in H_P, \\ +\infty, & h \notin H_P. \end{cases}$$

Then the following result is true.

Theorem 8.3 (Gaussian Large Deviation Principle) *For any open set $G \subset \mathscr{X}$ it is true that*

$$\liminf_{R \to \infty} \frac{\ln \mathbb{P}\{X \in RG\}}{R^2} \geq -J(G). \tag{8.1}$$

For any closed set $F \subset \mathscr{X}$ it is true that

$$\limsup_{R \to \infty} \frac{\ln \mathbb{P}\{X \in RF\}}{R^2} \leq -J(F), \tag{8.2}$$

where $J(A) = \inf_A I(\cdot)$.

The set $A \subset \mathscr{X}$ is called *regular*, if

$$\inf_{Cl(A)} I(\cdot) = \inf_A I(\cdot) = \inf_{Int(A)} I(\cdot).$$

Theorem 8.3 yields

$$\lim_{R \to \infty} \frac{\ln \mathbb{P}\{X \in RA\}}{R^2} = -J(A)$$

for any regular set.

Proof (of Theorem 8.3). Lower bound. Let G be an open set and $h \in G \cap H_P$. Then there exists a symmetric neighborhood of the origin V such that $h + V \subset G$. By using Proposition 5.1 we obtain

$$P(RG) \geq P(Rh + RV) \geq P(RV)\exp\{-R^2|h|^2_{H_P}/2\},$$

while

$$\lim_{R \to \infty} P(RV) = P(\mathscr{X}) = 1. \tag{8.3}$$

It follows that

$$\liminf_{R \to \infty} \frac{\ln \mathbb{P}\{X \in RG\}}{R^2} \geq \frac{-|h|^2_{H_P}}{2}.$$

By maximizing over h we obtain (8.1).

Upper bound. Let F be a closed set and $\rho < \inf_{h \in F} |h|_{H_P}$. The the ball $B := \{h \in H_P : |h|_{H_P} \leq \rho\}$ and the set F are disjoint. We use that B is a compact set. Since F

is closed, for any point $h \in B$ there exists a convex neighborhood of the origin V_h such that $h + V_h \cap F = \emptyset$. From the covering $\{h + V_h/2\}$ of the set B we extract a finite sub-covering $\{h_i + V_{h_i}/2\}$ and let

$$V = \bigcap_i (V_{h_i})/2.$$

Then

$$(B + V) \cap F = \emptyset. \tag{8.4}$$

Indeed, for any $h \in B$ there exists an i such that $h \in h_i + V_{h_i}/2$. Therefore, for any $v \in V$ we have

$$h + v \in h_i + V_{h_i}/2 + V \subset h_i + V_{h_i}/2 + V_{h_i}/2 = h_i + V_{h_i}.$$

Thus, $h + v \notin F$ and (8.4) follows. Now, by applying isoperimetric inequality we obtain

$$P(RF) \leq P(\mathcal{X} \setminus (RB + RV)) \leq 1 - \Phi\left(\Phi^{-1}(P(RV)) + R\rho\right).$$

It follows from (8.3) that for R large enough we have

$$\Phi^{-1}(P(RV)) \geq \Phi^{-1}(1/2) = 0,$$

hence

$$P(RF) \leq 1 - \Phi(R\rho).$$

It follows that

$$\limsup_{R \to \infty} \frac{\ln \mathbb{P}\{X \in RF\}}{R^2} \leq \limsup_{R \to \infty} \frac{\ln(1 - \Phi(R\rho))}{R^2} = \frac{-\rho^2}{2}.$$

It remains to pass to the limit $\rho \nearrow \inf_{h \in F} |h|_{H_P}$. Then $\frac{\rho^2}{2} \nearrow J(F)$, and we obtain (8.2). $\qquad\qquad\square$

Remark 8.1 Gaussian large deviation principle was independently established by Wentzell [181] and Freidlin [70] in Hilbert space. Even earlier, a particular case of Wiener measure was considered by Schilder [155].

8.3 Applications of Large Deviation Principle

Example 8.1 (Large deviations for maximum of Gaussian process) Let $X(t), t \in T$, be a centered Gaussian process with continuous sample paths on a compact T. Then

$$\lim_{R \to \infty} \frac{\ln \mathbb{P} (\max_T X \geq R)}{R^2} = \frac{-1}{2\sigma^2}, \tag{8.5}$$

$$\lim_{R \to \infty} \frac{\ln \mathbb{P} (\max_T |X| \geq R)}{R^2} = \frac{-1}{2\sigma^2}, \tag{8.6}$$

where

$$\sigma^2 = \max_T \mathbb{E} \, X(t)^2.$$

We apply Large Deviation Principle to the vector $X = (X(t))$ in space $\mathbb{C}(T)$ and to the set $A = \{x : \max_T x \geq 1\}$. Notice that A is a regular set. Indeed, A is closed, thus $J(A) = J(Cl(A))$. On the other hand, $Int(A) = \{x : \max_T x > 1\}$. Therefore, for any $h \in A$ and any $\varepsilon > 0$ we have $(1 + \varepsilon)h \in Int(A)$ and

$$I(h) = \lim_{\varepsilon \to 0} (1 + \varepsilon)^2 I(h) = \lim_{\varepsilon \to 0} I((1 + \varepsilon)h) \geq J(Int(A)).$$

It follows that

$$J(A) = \inf_{h \in A} I(h) \geq J(Int(A)).$$

The converse inequality $J(A) \leq J(Int(A))$ is obvious, thus the regularity of A is verified. It follows that the limit in (8.5) exists and is equal to $-J(A)$. Let us compute it.

For any $t \in T$ we let $\sigma_t^2 = \mathbb{E}X(t)^2$, $\delta_t x = x(t)$, and $h_t = \sigma_t^{-2} K\delta_t = \sigma_t^{-2} II^* \delta_t$. Then

$$|h_t|_{H_P}^2 = \sigma_t^{-4} ||I^* \delta_t||_{\mathscr{X}_P^*}^2 = \sigma_t^{-4} \sigma_t^2 = \sigma_t^{-2}$$

and

$$h_t(t) = (\delta_t, h) = (\delta_t, \sigma_t^{-2} II^* \delta_t) = \sigma_t^{-2} (I^* \delta_t, I^* \delta_t) = 1.$$

Therefore, $h_t \in A$ and

$$J(A) = \inf_{h \in A} I(h) \leq \inf_{t \in T} I(h_t) = \inf_{t \in T} \frac{|h_t|_{H_P}^2}{2} = \inf_{t \in T} \frac{\sigma_t^{-2}}{2} = \frac{1}{2\sigma^2}.$$

Conversely, let $h = Iz \in A$. Then we have for some $t \in T$

$$1 \leq h(t) = (\delta_t, h) = (\delta_t, Iz) = (I^* \delta_t, z) \leq \sigma_t ||z||_{\mathscr{X}_P^*}.$$

Therefore,

$$|h|_{H_P} = ||z||_{\mathscr{X}_P^*} \geq \inf_{t \in T} \sigma_t^{-1} = \sigma^{-1}.$$

It follows that $J(A) \geq \frac{1}{2\sigma^2}$. We obtain $J(A) = \frac{1}{2\sigma^2}$ and (8.5) is proved.

Exercise 8.1   Prove (8.6) by a similar reasoning.

Exercise 8.2   For any function $f \in \mathbb{C}[0, 1]$ define its continuity modulus by

$$\omega(f, u) := \sup_{|s-t| \leq u} |f(s) - f(t)|.$$

Let W be a Wiener process. Prove that

$$\lim_{R \to \infty} \frac{\ln \mathbb{P}(\omega(W, u) \geq R)}{R^2} = \frac{-1}{2u}, \quad 0 < u \leq 1. \tag{8.7}$$

Example 8.2   (Norm distribution of a Gaussian vector) Let X be a centered Gaussian vector in a separable Banach space $(\mathscr{X}, ||\cdot||)$. Let P be the distribution of X and let $K : \mathscr{X}^* \mapsto \mathscr{X}$ be the covariance operator of X. Then

$$\lim_{R \to \infty} \frac{\ln \mathbb{P}(||X|| \geq R)}{R^2} = \frac{-1}{2||K||}. \tag{8.8}$$

It is well known that the norm of a vector in Banach space can be written as

$$||x|| = \sup_{f \in S^*} (f, x), \quad x \in \mathscr{X}, \tag{8.9}$$

where S^* denotes the unit sphere in the dual space. Therefore, we have

$$||X|| = \sup_{f \in S^*} (f, X),$$

and the evaluations of the preceding example apply to functionals $f \in S^*$ used instead of times $t \in T$. We obtain

$$\lim_{R \to \infty} \frac{\ln \mathbb{P}(||X|| \geq R)}{R^2} = \frac{-1}{2\sigma^2},$$

where

$$\sigma^2 = \sup_{f \in S^*} \mathbb{E}(f, X)^2 = \sup_{f \in S^*} (f, Kf).$$

It is obvious that

$$\sigma^2 \leq \sup_{f \in S^*} ||f|| \, ||Kf|| \leq ||K||.$$

On the other hand,

$$\|K\| = \sup_{f \in S^*} \|Kf\| = \sup_{f,g \in S^*} (g, Kf) = \sup_{f,g \in S^*} \mathbb{E}(f, X)(g, X)$$

$$\leq \left\{ \sup_{f \in S^*} \mathbb{E}(f, X)^2 \sup_{g \in S^*} \mathbb{E}(g, X)^2 \right\}^{1/2} = \sigma^2.$$

Therefore, $\sigma^2 = \|K\|$, and (8.8) is proved.

Exercise 8.3 (Moments' equivalence for the norm of a Gaussian vector). By using Concentration Principle (applied to the supremum functional, cf. representation (8.9) and Example 6.1), as well as inequality (7.4), prove that for any $p \geq 1$ there exists a positive constant c_p such that for any Gaussian vector in a normed space we have

$$\mathbb{E}\|X\| \leq \left(\mathbb{E}\|X\|^p\right)^{1/p} \leq c_p \mathbb{E}\|X\|. \tag{8.10}$$

9 Functional Law of the Iterated Logarithm

9.1 Classical Law of the Iterated Logarithm

Recall two classical forms of the law of iterated logarithm (LIL)

- *for sums of i.i.d. random variables*. Let (X_j) be a sequence of i.i.d. random centered random variables with unit variance. Let

$$S_n = \sum_{j=1}^{n} X_j.$$

Then Hartman–Wintner LIL [81] asserts that almost surely (a.s.)

$$\limsup_{n \to \infty} \frac{S_n}{\sqrt{2n \ln \ln n}} = 1, \quad \liminf_{n \to \infty} \frac{S_n}{\sqrt{2n \ln \ln n}} = -1. \tag{9.1}$$

- *for Wiener process*. Khinchin's LIL [32, 94] asserts that a.s.

$$\limsup_{T \to \infty} \frac{W(T)}{\sqrt{2T \ln \ln T}} = 1, \quad \liminf_{T \to \infty} \frac{W(T)}{\sqrt{2T \ln \ln T}} = -1.$$

The similarity of assertions is explained by the fact that for $T = n$ one can represent $W(T)$ as a sum of i.i.d. increments $W(j) - W(j - 1)$.

Let us stress that LIL is a form of investigation *of large deviations* for the corresponding variables. For example, according to the central limit theorem, a typical value for S_n is of order $\sqrt{n}$, while LIL deals with the values of larger order $\sqrt{n \ln \ln n}$. The same observation is true for Wiener process.

LIL was initially discovered by Khinchin [90] in the context of Bernoulli scheme. Subsequent developments of LIL marked by the works of Kolmogorov, Lévy, Erdös, Feller, Petrov, Ledoux etc are very rich, see [19, 66, 107, 145] and references therein.

9.2 Functional Law of the Iterated Logarithm

We pass now to the study of the *functional LIL*, denoted further by FLIL, for Wiener process. Unlike the classical LIL, the FLIL takes into account not only the value $W(T)$, but the entire sample path $W(t), 0 \leq t \leq T$. For any $T > 3$ let us introduce two families of random elements of the space $\mathbb{C}[0, 1]$ by

$$X_T(s) = \frac{W(sT)}{\sqrt{T}}, \quad 0 \leq s \leq 1,$$

$$Y_T(s) = \frac{W(sT)}{\sqrt{2T \ln \ln T}} = \frac{X_T(s)}{\sqrt{2 \ln \ln T}}, \quad 0 \leq s \leq 1. \tag{9.2}$$

Notice that both of them contain, in a compressed form, the whole sample path $W(t), 0 \leq t \leq T$. Moreover,

$$Y_T(1) = \frac{W(T)}{\sqrt{2T \ln \ln T}}.$$

Therefore, the study of $Y(\cdot)$ is a natural task from Khinchin's LIL point of view.

By self-similarity, X_T itself is a Wiener process for each T.

In what follows we handle a *convergence to a set*. Recall the corresponding definition. Let $(Y_T)_{T \geq T_0}$ be a family of elements of a metric space $(\mathscr{X}, \rho)$. A compact subset $K \subset \mathscr{X}$ is called a *limit set* for Y_T, as $T \to \infty$, if two conditions are satisfied

1. $\lim_{T \to \infty} \inf_{h \in K} \rho(Y_T, h) = 0$.
2. for any $h \in K$ it is true that $\liminf_{T \to \infty} \rho(Y_T, h) = 0$.

In this case we write $Y_T \hookrightarrow K$. The first condition tells that for large T the element Y_T is close to *some* element of the compact set K. The second condition tells that any neighborhood of *any* element of K is visited by the family (Y_T) at arbitrary large times T.

If $Y_T \hookrightarrow K$, then for any continuous functional $g : \mathscr{X} \mapsto \mathbb{R}$ it is true that

$$\limsup_{T \to \infty} g(Y_T) = \sup_{h \in K} g(h), \quad \liminf_{T \to \infty} g(Y_T) = \inf_{h \in K} g(h). \tag{9.3}$$

Therefore, the investigation of limit behavior of a functional reduces to solving of an extremal problem on the limit compact set. For proving (9.3) notice that the second convergence property and the continuity of g imply that for any $h \in K$

$$L := \limsup_{T \to \infty} g(Y_T) \geq g(h),$$

and, optimizing over h, we find that

$$L \geq \sup_{h \in K} g(h).$$

Conversely, take a sequence $T_n \to \infty$ such that $\lim_{n \to \infty} g(Y_{T_n}) = L$. By the first convergence property, there is a sequence $h_n \in K$ such that

$$\lim_{n \to \infty} \rho(Y_{T_n}, h_n) = 0.$$

By using the compactness of K, we may extract a convergent subsequence $h_{n_j} \to h \in K$. Then $Y_{T_{n_j}} \to h$ and

$$L = \lim_{j \to \infty} g(Y_{T_{n_j}}) = g(h),$$

whence $L \leq \sup_{h \in K} g(h)$. The first assertion in (9.3) is proved. The second one follows by replacing g with $-g$. □

In FLIL, the limiting set is the unit ball of the kernel of Wiener measure (dispersion ellipsoid), i.e. the set

$$K = \left\{ h : |h|_{Hp} \leq 1 \right\} = \left\{ h : h \in AC[0,1] : h(0) = 0, \int_0^1 h'(s)^2 ds \leq 1 \right\}. \quad (9.4)$$

In FLIL context K is often called *Strassen ball*.

Theorem 9.1 (FLIL (or Strassen FLIL) for Wiener process [161]) *Let the family of processes* $(Y_T)_{T \geq 3}$ *be given by (9.2), and let the set K be given by (9.4). Then*

$$Y_T \hookrightarrow K \quad a.s.$$

Proof (of Theorem 9.1). It is long but very instructive. The leading idea is an exponential blocking: we mainly follow the behavior of Y_T along exponentially increasing sequences $T_n = \gamma^n$, where parameter $\gamma > 1$ will be chosen according to current needs. Introduce an appropriate notation. Let $\gamma > 1$. We represent Y_T as

$$Y_T = \widehat{Y}_T + Y_T^0,$$

where $\widehat{Y}_T(s) := Y_T\left(\min(s, \gamma^{-1})\right)$ is a function $Y_T(\cdot)$ stopped at time $\gamma^{-1} < 1$. Accordingly, $Y_T^0 := Y_T - \widehat{Y}_T$. It is easy to see that a function

$$Y_T^0(s) = \begin{cases} 0, & 0 \leq s \leq \gamma^{-1}, \\ (2T \ln \ln T)^{-1/2} \left(W(sT) - W(\gamma^{-1}T) \right), & \gamma^{-1} \leq s \leq 1, \end{cases}$$

is completely determined by the increments of W on the interval $[\gamma^{-1}T, T]$. Therefore, when we consider a sequence of times $T_n = \gamma^n$, the random functions $Y_{T_n}^0$ are independent. Let $\rho(x, y) = \|x - y\|$ denote the uniform distance between functions.

The technical part of the proof consists of four statements about series convergence.

1. For any $\gamma > 1$ and $\varepsilon > 0$ it is true that

$$\sum_{n=1}^{\infty} \mathbb{P}\left(\inf_{h \in K} \rho(Y_{T_n}, h) > \varepsilon \right) < \infty. \quad (9.5)$$

2. For any $\gamma > 1$ there exists $\varepsilon_1 = \varepsilon_1(\gamma) > 0$ such that

$$\sum_{n=1}^{\infty} \mathbb{P}\left(\sup_{T_n \leq T \leq T_{n+1}} \rho(Y_T, Y_{T_n}) > \varepsilon_1 \right) < \infty \tag{9.6}$$

and $\lim_{\gamma \searrow 0} \varepsilon_1(\gamma) = 0$.

3. For any $h \in K$, $\gamma > 1$ and $\varepsilon > 0$ it is true that

$$\sum_{n=1}^{\infty} \mathbb{P}\left(\rho(Y_{T_n}, h) < \varepsilon \right) = \infty. \tag{9.7}$$

4. For any $\gamma > 1$ there exists $\varepsilon_2 = \varepsilon_2(\gamma) > 0$ such that

$$\sum_{n=1}^{\infty} \mathbb{P}\left(\| \widehat{Y}_{T_n} \| > \varepsilon_2 \right) < \infty \tag{9.8}$$

and $\lim_{\gamma \to \infty} \varepsilon_2(\gamma) = 0$.

By using first two claims we will prove that Y_T approaches K, as T is large; by using two remaining claims we will prove that every element of K is a limit point.

From now on, all statements we will do in the proof hold with probability one. By Borel–Cantelli Lemma, (9.5) proves that for any $\gamma > 1$ it is true that

$$\limsup_{n \to \infty} \inf_{h \in K} \rho(Y_{T_n}, h) = 0,$$

while (9.6) implies

$$\limsup_{n \to \infty} \sup_{T_n \leq T \leq T_{n+1}} \rho(Y_T, Y_{T_n}) \leq \varepsilon_1.$$

By triangle inequality

$$\inf_{h \in K} \rho(Y_T, h) \leq \inf_{h \in K} \rho(Y_{T_n}, h) + \rho(Y_T, Y_{T_n}),$$

whence

$$\limsup_{T \to \infty} \inf_{h \in K} \rho(Y_T, h) \leq \varepsilon_1.$$

Finally, letting $\gamma \searrow 1$, we obtain

$$\limsup_{T \to \infty} \inf_{h \in K} \rho(Y_T, h) \leq \lim_{\gamma \searrow 1} \varepsilon_1(\gamma) = 0.$$

Let us fix an $h \in K$ and a $\gamma > 1$. A difference of a divergent series and a convergent one is a divergent series. Therefore, (9.7) and (9.8) imply

$$\sum_{n=1}^{\infty} \mathbb{P}\left(\rho(Y_{T_n}, h) \leq \varepsilon, \left\|\widehat{Y}_{T_n}\right\| \leq \varepsilon_2\right)$$

$$\geq \sum_{n=1}^{\infty} \left(\mathbb{P}\left(\rho(Y_{T_n}, h) \leq \varepsilon\right) - \mathbb{P}\left(\left\|\widehat{Y}_{T_n}\right\| > \varepsilon_2\right)\right) = \infty.$$

By letting $\varepsilon = \varepsilon_2$ and assuming $\rho(Y_{T_n}, h) < \varepsilon_2$, $\left\|\widehat{Y}_{T_n}\right\| \leq \varepsilon_2$, we derive from triangle inequality

$$\rho(Y_{T_n}^0, h) \leq \rho(Y_{T_n}, h) + \rho(Y_{T_n}^0, Y_{T_n}) = \rho(Y_{T_n}, h) + \left\|\widehat{Y}_{T_n}\right\| \leq 2\varepsilon_2.$$

Therefore,

$$\sum_{n=1}^{\infty} \mathbb{P}\left(\rho(Y_{T_n}^0, h) \leq 2\varepsilon_2\right) = \infty.$$

Since the random elements $Y_{T_n}^0$ are independent, Borel–Cantelli Lemma yields

$$\liminf_{n \to \infty} \rho(Y_{T_n}^0, h) \leq 2\varepsilon_2.$$

On the other hand, (9.8) and Borel–Cantelli Lemma yield

$$\limsup_{n \to \infty} \left\|\widehat{Y}_{T_n}\right\| \leq \varepsilon_2.$$

By triangle inequality,

$$\rho(Y_{T_n}, h) \leq \rho(Y_{T_n}^0, h) + \rho(Y_{T_n}^0, Y_{T_n}),$$

therefore,

$$\liminf_{n \to \infty} \rho(Y_{T_n}, h) \leq \liminf_{n \to \infty} \rho(Y_{T_n}^0, h) + \limsup_{n \to \infty} \left\|\widehat{Y}_{T_n}\right\| \leq 3\varepsilon_2.$$

Finally,

$$\liminf_{T \to \infty} \rho(Y_T, h) \leq \liminf_{n \to \infty} \rho(Y_{T_n}, h) \leq 3\varepsilon_2,$$

and letting $\gamma \to \infty$, we obtain

$$\liminf_{T \to \infty} \rho(Y_T, h) \leq 3 \lim_{\gamma \to \infty} \varepsilon_2(\gamma) = 0.$$

It remains to check the claims (9.5)–(9.8).

Let us prove (9.5). Let denote $L_T = 2 \ln \ln T$ and $\mathbb{U} := \{x \in \mathbb{C}[0, 1] : \|x\| \leq 1\}$ the unit ball in $\mathbb{C}[0, 1]$. Since

$$\lim_{r \to \infty} \mathbb{P}(W \in r\mathbb{U}) = 1, \tag{9.9}$$

isoperimetric inequality (6.8) implies

$$\mathbb{P}\left(\inf_{h \in K} \rho(Y_{T_n}, h) > \varepsilon\right) = \mathbb{P}\left(Y_{T_n} \notin K + \varepsilon\mathbb{U}\right) = \mathbb{P}\left(\frac{W}{\sqrt{L_{T_n}}} \notin K + \varepsilon\mathbb{U}\right)$$

$$= \mathbb{P}\left(W \notin \sqrt{L_{T_n}}K + \varepsilon\sqrt{L_{T_n}}\mathbb{U}\right) \le \widehat{\Phi}\left(\Phi^{-1}\left(\mathbb{P}(W \in \varepsilon\sqrt{L_{T_n}}\mathbb{U}) + \sqrt{L_{T_n}}\right)\right)$$

$$\le \widehat{\Phi}\left(1 + \sqrt{L_{T_n}}\right) \le \exp\left\{\frac{-(1 + \sqrt{L_{T_n}})^2}{2}\right\} \le \exp\left\{\frac{-L_{T_n}}{2} - \sqrt{L_{T_n}}\right\},$$

whenever n is sufficiently large. By applying a crude estimate $\exp(-x) \le cx^{-4}$ to $x = \sqrt{L_{T_n}}$, we obtain

$$\mathbb{P}\left(\inf_{h \in K} \rho(Y_{T_n}, h) > \varepsilon\right) \le \exp\left\{\frac{-L_{T_n}}{2} - \sqrt{L_{T_n}}\right\} \le c\,L_{T_n}^{-2}\exp\left\{\frac{-L_{T_n}}{2}\right\}$$

$$= \frac{c}{(2(\ln\ln\gamma + \ln n))^2}\frac{1}{\ln T_n} = \frac{c}{(2(\ln\ln\gamma + \ln n))^2}\frac{1}{\ln\gamma\,n},$$

and this leads to a convergent series. Relation (9.5) is proved.

Let us prove (9.7). Let $\varepsilon > 0$ and $h \in K$. By using (9.9) and (5.3) we obtain

$$\mathbb{P}\left(\rho(Y_{T_n}, h) \le \varepsilon\right) = \mathbb{P}\left(Y_{T_n} \in h + \varepsilon\mathbb{U}\right) = \mathbb{P}\left(\frac{W}{\sqrt{L_{T_n}}} \in h + \varepsilon\mathbb{U}\right)$$

$$= \mathbb{P}\left(W \in h\sqrt{L_{T_n}} + \varepsilon\sqrt{L_{T_n}}\mathbb{U}\right)$$

$$\ge \exp\left\{-\frac{L_{T_n}^2\,|h|_{Hp}^2}{2}\right\}\mathbb{P}\left(W \in \varepsilon\sqrt{L_{T_n}}\mathbb{U}\right)$$

$$\ge (\ln(T_n))^{-|h|_{Hp}^2}\frac{1}{2} = \frac{1}{2}(\ln\gamma\,n)^{-|h|_{Hp}^2},$$

whenever n is sufficiently large. Since $|h|_{Hp} \le 1$, the latter expressions form a divergent series and the claim (9.7) is proved.

Let us prove (9.8). Let $\varepsilon > 0$ and $\gamma > 1$. By using self-similarity of Wiener process, we have

$$\mathbb{P}\left(\|\widehat{Y}_{T_n}\| > \varepsilon\right) = \mathbb{P}\left(\frac{\|W(\min(\cdot, \gamma^{-1}))\|}{\sqrt{L_{T_n}}} > \varepsilon\right)$$

$$= \mathbb{P}\left(\max_{0 \le s \le \gamma^{-1}} |W(s)| > \varepsilon\sqrt{L_{T_n}}\right)$$

$$= \mathbb{P}\left(\gamma^{-1/2}\max_{0 \le s \le 1} |W(s)| > \varepsilon\sqrt{L_{T_n}}\right) = \mathbb{P}\left(\max_{0 \le s \le 1} |W(s)| > \varepsilon\sqrt{\gamma L_{T_n}}\right).$$

By applying Large Deviation Principle for maxima (8.6), we have

$$\mathbb{P}\left(\max_{0\leq s\leq 1}|W(s)| > \varepsilon\sqrt{\gamma L T_n}\right) \leq \exp\left\{\frac{-\varepsilon^2\gamma L T_n}{3}\right\}$$

$$= (\ln(T_n))^{-2\varepsilon^2\gamma/3} = (\ln \gamma n)^{-2\varepsilon^2\gamma/3},$$

whenever n is large enough. Such quantities form a convergent series if we choose $\varepsilon := \varepsilon_2(\gamma) = 2\gamma^{-1/2}$. This choice provides both (9.8) and $\lim_{\gamma\to\infty}\varepsilon_2(\gamma) = 0$.

Let us prove (9.6). Let $\varepsilon > 0$, $\gamma > 1$ and $T \in [T_n, T_{n+1}]$. The definition of Y_T implies

$$Y_T(s) = \frac{W(sT)}{\sqrt{T L_T}} = \frac{\sqrt{T_{n+1}}}{\sqrt{T L_T}} X_{T_{n+1}}\left(\frac{sT}{T_{n+1}}\right).$$

Therefore,

$$\rho(Y_T, Y_{T_{n+1}}) = \sup_{0\leq s\leq 1}\left|\frac{\sqrt{T_{n+1}}}{\sqrt{T L_T}} X_{T_{n+1}}\left(\frac{sT}{T_{n+1}}\right) - \frac{1}{\sqrt{L_{T_{n+1}}}} X_{T_{n+1}}(s)\right|$$

$$\leq \sup_{0\leq s\leq 1}\Delta_1(s) + \sup_{0\leq s\leq 1}\Delta_2(s),$$

where

$$\Delta_1(s) = \frac{\sqrt{T_{n+1}}}{\sqrt{T L_T}}\left|X_{T_{n+1}}\left(\frac{sT}{T_{n+1}}\right) - X_{T_{n+1}}(s)\right|;$$

$$\Delta_2(s) = \left|\frac{\sqrt{T_{n+1}}}{\sqrt{T L_T}} - \frac{1}{\sqrt{L_{T_{n+1}}}}\right| \cdot \left|X_{T_{n+1}}(s)\right|.$$

For the first term, we use a large deviation estimate for modulus of continuity of Wiener process (8.7). For large n, we obtain

$$\mathbb{P}\left\{\sup_{0\leq s\leq 1}\Delta_1(s) \geq \varepsilon\right\} = \mathbb{P}\left\{\frac{\sqrt{T_{n+1}}}{\sqrt{T L_T}}\sup_{0\leq s\leq 1}\left|W\left(\frac{sT}{T_{n+1}}\right) - W(s)\right| \geq \varepsilon\right\}$$

$$\leq \mathbb{P}\left\{\omega\left(W, \left(1 - \frac{1}{\gamma}\right)\right) \geq \frac{\varepsilon\sqrt{T L_T}}{\sqrt{T_{n+1}}}\right\}$$

$$= \mathbb{P}\left\{\omega\left(W, \left(1 - \frac{1}{\gamma}\right)\right) \geq \frac{\varepsilon\sqrt{L T_n}}{\sqrt{\gamma}}\right\}$$

$$\leq \exp\left\{-\frac{\varepsilon^2 L T_n}{3\gamma}\left(1 - \frac{1}{\gamma}\right)^{-1}\right\} = ((\ln \gamma)n)^{-\frac{2\varepsilon^2}{3\gamma}\left(1-\frac{1}{\gamma}\right)^{-1}}.$$

If $\varepsilon \geq 2(\gamma - 1)^{1/2}$, then the corresponding series converges. The second term is handled similarly and even simpler. We obtain

$$\mathbb{P}\left\{\sup_{0\le s\le 1}\Delta_2(s)\ge\varepsilon\right\}\le\mathbb{P}\left\{||W||\ge\frac{c\varepsilon\sqrt{L_{T_n}}}{\left(1-\frac{1}{\gamma}\right)}\right\}$$

$$\le\exp\left\{-\frac{c^2\varepsilon^2}{3}L_{T_n}\left(1-\frac{1}{\gamma}\right)^{-2}\right\}=((\ln\gamma)n)^{-\frac{2c^2\varepsilon^2}{3}\left(1-\frac{1}{\gamma}\right)^{-2}}.$$

If $\varepsilon\ge 2c^{-1}\left(1-\frac{1}{\gamma}\right)$, then the corresponding series converges. It remains to choose

$$\varepsilon_1(\gamma)=2\max\left\{2(\gamma-1)^{1/2};2c^{-1}\left(1-\frac{1}{\gamma}\right)\right\},$$

and (9.6) is proved, along with $\lim_{\gamma\searrow 1}\varepsilon_1(\gamma)=0$.

Exercise 9.1 Let W be a Wiener process. Prove that the upper limit

$$\limsup_{T\to\infty}\frac{\int_0^T W(s)^2\,ds}{T^2\ln\ln T}$$

is non-random (with probability one) and calculate it.

9.3 Some Extensions

Convergence rate

It is interesting to find out how fast is the convergence rate in FLIL. Since FLIL is a pair of assertions, the convergence rate should be evaluated separately for each of them. As for approaching of Y_T to the set K, Talagrand and Grill [79, 167] independently proved that for some c_1, c_2

$$\frac{c_1}{(\ln\ln T)^{2/3}}\le\inf_{h\in K}\rho(Y_T,h)\le\frac{c_2}{(\ln\ln T)^{2/3}},$$

holds almost surely, whenever n is sufficiently large. As for the approximation of specific element $h\in K$, one can prove the following. If h is an *interior* element of K, i.e. $|h|_{H_P}<1$, then the convergence rate is of order $(\ln\ln T)^{-1}$. More precisely, the results of Csáki and de Acosta [42, 44] show that

$$\liminf_{T\to\infty}(||Y_T-h||\ln\ln T)=\frac{\pi}{4\sqrt{1-|h|_{H_P}^2}}.$$

For the *boundary* elements, $|h|_{H_P}=1$, the approximation rate depends on h. The rate is always slower than $(\ln\ln T)^{-1}$ but at least as fast as $(\ln\ln T)^{-2/3}$.

Exercise 9.2 (Chung's FLIL, [36, 85, 86]) Prove that

$$\liminf_{T \to \infty} \left(\frac{\ln \ln T}{T} \right)^{1/2} \sup_{0 \le t \le T} |W(t)| = \frac{\pi}{\sqrt{8}} \quad \text{a.s.}$$

Other Norms

In FLIL, the space $\mathbb{C}[0, 1]$ can be replaced [10, 45, 46] with other normed spaces that contain sample paths of Wiener process with probability one, for example, $L_p[0, 1]$ or the space of α-Hölder functions equipped with the norm

$$\|x\|_\alpha = \sup_{\substack{s,t \in [0,1] \\ s \ne t}} \frac{|x(s) - x(t)|}{|s - t|^\alpha}, \quad 0 < \alpha < \frac{1}{2}.$$

Stronger the norm is, larger is the corresponding distance, larger is the class of continuous functionals, thus stronger are the results provided by FLIL. The convergence rate in FLIL depends on the choice of the norm.

Multivariate Process

We can replace W with its multivariate analogue, i.e. a vector-valued process

$$W(t) = \left(W^{(1)}(t), \ldots, W^{(n)}(t) \right) \in \mathbb{R}^n,$$

where $(W^{(j)}(t))_{j=1}^n$ are independent scalar Wiener processes. Keeping the same definition of Y_T given in (9.2), we obtain convergence to a set

$$K^n = \left\{ h \in \mathbb{C}_0([0, 1], \mathbb{R}^n) : h_j(\cdot) \in AC[0, 1], \int_0^1 \left(\sum_{j=1}^n h'_j(s)^2 \right) ds \le 1 \right\}.$$

Exercise 9.3 By using (9.3), find the limit

$$\limsup_{T \to \infty} \frac{\|W(t)\|}{\sqrt{T \ln \ln T}}$$

for n-dimensional Wiener process.

Multi-parametric process

Consider a Brownian sheet $W(t), t \in \mathbb{R}_+^d$ defined in Example 2.7. In view of its self-similarity, it is natural to consider random elements

$$Y_T(s) = \frac{W(Ts)}{\sqrt{2T^d \ln \ln T}}, \quad T > 3, s \in [0, 1]^d,$$

in $\mathbb{C}([0, 1]^d)$ and prove that their limit set is the unit ball of the kernel of W, cf. Example 4.7 (see [14, 143]).

Fractional Brownian Motion

Let $W^{(\alpha)}(t)$ be an α-fractional Brownian motion. Self-similarity of $W^{(\alpha)}$ means that a process

$$X_T(s) = \frac{W^{(\alpha)}(sT)}{T^{\alpha/2}}$$

also is an fBm. Therefore, we have to deal with

$$Y_T(s) = \frac{W^{(\alpha)}(sT)}{\sqrt{2T^\alpha \ln \ln T}} = \frac{X_T(s)}{\sqrt{2 \ln \ln T}}, \quad 0 \le s \le 1.$$

FLIL for fBm asserts that $Y_T \hookrightarrow K^{(\alpha)}$, where the limit set $K^{(\alpha)}$ coincides with the unit ball of the kernel of $W^{(\alpha)}$, cf. Example 4.6.

9.4 Strong Invariance Principle

In order to derive an FLIL for random walks extending Hartman–Wintner theorem, it is sufficient to check that the sums are close to the values of an appropriate Wiener process. Such statements are referred to as *Strong Invariance Principles*. Lat us recall the most important results of this kind. The general scheme is as follows. Let (X_j) be an i.i.d. sequence of centered random variables with unit variance defined on a common probability space. Assume that a sequence of variables $(\tilde{X}_j)$ equidistributed with (X_j) and a Wiener process W are jointly defined on some other probability space. Let

$$\tilde{S}_k = \sum_{j=1}^{k} \tilde{X}_j$$

and

$$\Delta_n = \max_{1 \le k \le n} |\tilde{S}_k - W(k)|.$$

Strong Invariance Principle asserts that under certain moment restrictions imposed on the common distribution of variables (X_j) a construction of $(\tilde{X}_j)$ and W with given decay of Δ_n is possible. Let us mention some examples, moving from stronger assumptions to weaker ones.

- Komlós-Major-Tusnády Strong Invariance Principle. (KMT-construction, [95, 96]). If $\mathbb{E} \exp(c|X_j|) < \infty$ holds for some $c > 0$, then

$$\limsup_{n \to \infty} \frac{\Delta_n}{\ln n} < \infty \quad \text{a.s.}$$

- Sakhanenko Strong Invariance Principle [152]. If $\mathbb{E}|X_j|^p < \infty$ for some $p > 2$, then

$$\lim_{n \to \infty} \frac{\Delta_n}{n^{1/p}} = 0 \quad \text{a.s.}$$

- Strassen Strong Invariance Principle [161]. The weakest assumption $\mathbb{E}|X_j|^2 = 1$ (that was anyway already imposed in the beginning of subsection) yields

$$\lim_{n \to \infty} \frac{\Delta_n}{(n \ln \ln n)^{1/2}} = 0 \quad \text{a.s.} \tag{9.10}$$

This estimate can not be ameliorated without further moment assumptions. However, the following result shows that a slightly better approximation rate is available, if we change a bit the approximating term.

- Major Strong Invariance Principle [132]. Under the same assumptions we have

$$\lim_{n \to \infty} \frac{\tilde{\Delta}_n}{n^{1/2}} = 0 \quad \text{a.s.},$$

where

$$\tilde{\Delta}_n = \max_{1 \le k \le n} |\tilde{S}_k - W(\tilde{k})|,$$

and

$$\tilde{k} = \sum_{j=1}^{k} \mathbb{V}ar\left(X_j \cdot \mathbf{1}_{|X_j| \le j}\right)$$

satisfies $\tilde{k} \le k$ and $\lim_{k \to \infty} \frac{\tilde{k}}{k} = 1$.

9.5 FLIL for Random Walk

Let (X_j) and S_k be the same as in Hartman–Wintner Theorem (9.1). Define the scaled sample paths of random walk $Z_n(s)$, $0 \le s \le 1$, by

$$Z_n\left(\frac{k}{n}\right) = \begin{cases} 0, & k = 0, \\ \frac{S_k}{\sqrt{2n \ln \ln n}}, & 1 \le k \le n, \end{cases} \tag{9.11}$$

using linear interpolation between the points $\frac{k}{n}$.

Theorem 9.2 (FLIL (or *Strassen law*) for random walks [161]. *Let a family* $(Z_n)_{n\geq 3}$ *be given by* (9.11), *and let K be defined in* (9.4). *Then*

$$Z_n \hookrightarrow K \quad \text{a.s.}$$

Proof (*of Theorem 9.2*). Let us construct a sequence $(\tilde{X}_j)$ equidistributed with (X_j) and a Wiener process W, as in Strassen's Strong Invariance Principle. Let $\tilde{Z}_n$ be a sequence, analogous to Z_n, constructed by using $(\tilde{X}_j)$. Since (X_j) and $(\tilde{X}_j)$ are equidistributed, the limit sets for (Z_n) and $(\tilde{Z}_n)$ coincide. Therefore, it is enough to show that $\tilde{Z}_n \hookrightarrow K$. Let Y_n be the scaled sample paths of W, constructed as in (9.2) with $T=n$, and let $\tilde{Y}_n$ be their linear interpolations over the grid $\{\frac{k}{n}, 0 \leq k \leq n\}$. Then

$$\rho(\tilde{Z}_n, \tilde{Y}_n) = \max_{0\leq s\leq 1} |\tilde{Z}_n(s) - \tilde{Y}_n(s)| = \max_{0\leq k\leq n} |\tilde{Z}_n(\frac{k}{n}) - \tilde{Y}_n(\frac{k}{n})|$$

$$= \max_{0\leq k\leq n} |\tilde{Z}_n(\frac{k}{n}) - Y_n(\frac{k}{n})| = \max_{0\leq k\leq n} \frac{|\tilde{S}_k - W(k)|}{\sqrt{2n \ln \ln n}}$$

$$= \frac{\Delta_n}{\sqrt{2n \ln \ln n}}.$$

From (9.10) it follows that $\lim_{n\to\infty} \rho(\tilde{Z}_n, \tilde{Y}_n) = 0$. On the other hand, on each interval $[\frac{k}{n}, \frac{k+1}{n}]$ the interpolation error $Y_n - \tilde{Y}_n$ is a scaled copy of a Brownian bridge. Moreover, these copies are independent. It is easy to conclude that $\lim_{n\to\infty} \rho(Y_n, \tilde{Y}_n) = 0$. By triangle inequality,

$$\lim_{n\to\infty} \rho(Y_n, \tilde{Z}_n) = 0.$$

Hence, $Y_n \hookrightarrow K$ yields $\tilde{Z}_n \hookrightarrow K$. $\square$

Remark 9.1 Hartman–Wintner Theorem (9.1) follows from Theorem 9.2 by applying (9.3) to the functional $g(x) = x(1)$.

10 Metric Entropy and Sample Path Properties

10.1 Basic Definitions

Let (T, ρ) be a metric space. Define a *covering number* $N(\varepsilon)$ as a minimal number of sets in a covering of T by subsets of diameter not exceeding ε. Then $N(\cdot)$ is a non-increasing function and $N(0+) < \infty$ iff T is a finite set and $N(\varepsilon) < \infty$ for all $\varepsilon > 0$ iff T is totally bounded. The quantity $H(\varepsilon) = \ln N(\varepsilon)$ is called *metric entropy* of the space T.

Define a *packing number* $M(\varepsilon)$ as a maximal number of points in T such that their pairwise distances exceed ε. Then $M(\cdot)$ is a non-increasing function and $M(0+) < \infty$

iff T is a finite set and $M(\varepsilon) < \infty$ for all $\varepsilon > 0$ iff T is totally bounded. The quantity $\mathscr{C}(\varepsilon) = \ln M(\varepsilon)$ is called *metric capacity* of the space T.

The following proposition shows that N and M basically describe the same property.

Proposition 10.1 *For any space (T, ρ) and for any $\varepsilon > 0$ it is true that*

$$N(2\varepsilon) \le M(\varepsilon) \le N(\varepsilon) \tag{10.1}$$

and

$$H(2\varepsilon) \le \mathscr{C}(\varepsilon) \le H(\varepsilon). \tag{10.2}$$

Proof. Take a configuration of $M(\varepsilon)$ points such that their pairwise distances exceed ε. Obviously, one can not add a point and keep this property. This means that the balls of radius ε centered at configuration points cover T. Since the diameter of any ball does not exceed 2ε, we obtain $N(2\varepsilon) \le M(\varepsilon)$. On the other hand, consider a covering of T that consists of $N(\varepsilon)$ subsets of diameter not exceeding ε and any configuration of points such that their pairwise distances exceed ε. Clearly, each covering element contains at most one configuration point. Therefore, the number of points in the configuration does not exceed $N(\varepsilon)$. Therefore, $M(\varepsilon) \le N(\varepsilon)$ and inequality (10.1) is proved. Inequality (10.2) follows from (10.1) by taking logarithms. □

Let now $X(t), t \in T$, be a Gaussian random process. We can introduce a *natural distance* on T generated by a process X, as

$$\rho(s, t)^2 := \mathbb{E}|X(s) - X(t)|^2.$$

Strictly speaking, ρ is not a distance because it is possible that $\rho(s, t) = 0$ for some $s \ne t$. This circumstance, however, does not affect the introduced entropy characteristics and their properties. An adept of absolute rigor may reduce the picture to a true metric space by identifying the points situated within zero distance.

In the following, $N, M, H, \mathscr{C}$ denote the quantities corresponding to the space (T, ρ). They will be used for investigation of properties of the process X.

10.2 Upper Bounds

Our main bound will be preceded by an important technical result.

Lemma 10.1 *Let $(X_j)_{1 \le j \le N}$ be centered Gaussian random variables and let $\sigma > 0$ satisfy $\max_{j \le N} \mathbb{E}X_j^2 \le \sigma^2$. Then*

$$\mathbb{E} \max_{1 \le j \le N} X_j \le \sqrt{2 \ln N}\,\sigma. \tag{10.3}$$

Proof. We evaluate the Laplace transform: for any λ it is true that

$$\mathbb{E}\exp\left\{\lambda\max_{1\le j\le N}X_j\right\}\le\mathbb{E}\left(\sum_{1\le j\le N}\exp\left\{\lambda X_j\right\}\right)$$

$$\le\sum_{1\le j\le N}\exp\left\{\lambda^2(\mathbb{E}X_j^2)/2\right\}$$

$$\le N\exp\left\{\lambda^2\sigma^2/2\right\}.$$

By Jensen inequality,

$$\lambda\,\mathbb{E}\max_{1\le j\le N}X_j\le\ln\mathbb{E}\exp\left\{\lambda\max_{1\le j\le N}X_j\right\}\le\ln N+\lambda^2\sigma^2/2.$$

By letting here $\lambda=\sqrt{2\ln N}\sigma^{-1}$, we obtain (10.3). $\qquad\qquad\square$

An integral based on the metric entropy,

$$\mathscr{D}(u)=\int_0^u\sqrt{H(\varepsilon)}\mathrm{d}\varepsilon$$

is called *Dudley integral* of a process X.

Theorem 10.1 (Dudley upper bound [52]) *For any centered Gaussian process* $X(t), t\in T$, *it is true that*

$$\mathbb{E}\sup_{t\in T}X(t)\le 4\sqrt{2}\mathscr{D}(\sigma/2),$$

where $\sigma^2=\sup_{t\in T}\mathbb{E}X(t)^2$.

Proof. The following evaluation scheme is often called *chaining scheme*. Let

$$\varepsilon_j=\sigma\cdot 2^{-j},\quad j=1,2,\dots$$

For any ε_j let us take a smallest covering of T by subsets of diameter not exceeding ε_j. Take a point in each element of the covering. Let denote S_j the set composed of $N(\varepsilon_j)$ chosen points. Let us fix a mapping $\pi_j:T\mapsto S_j$ such that

$$\rho(x,\pi_j(x))\le\varepsilon_j,\quad\forall x\in T.$$

Then it is true that

$$\sup_{t\in S_j}X(t)\le\sup_{t\in S_{j-1}}X(t)+\sup_{t\in S_j}(X(t)-X(\pi_{j-1}(t))).$$

Taking expectations and applying Lemma 10.1 to the last term, we obtain

$$\mathbb{E}\sup_{t\in S_j}X(t)\le\mathbb{E}\sup_{t\in S_{j-1}}X(t)+\sqrt{2H(\varepsilon_j)}\varepsilon_{j-1}.$$

Moreover, by applying the same lemma, we obtain

$$\mathbb{E} \sup_{t \in S_1} X(t) \le \sqrt{2H(\varepsilon_1)}\sigma.$$

Applying induction and using monotonicity of the function $H(\cdot)$, we infer that for each $j \ge 1$ it is true that

$$\mathbb{E} \sup_{t \in S_j} X(t) \le \sqrt{2H(\varepsilon_1)}\sigma + \sum_{k=2}^{j} \sqrt{2H(\varepsilon_k)}\varepsilon_{k-1}$$

$$= 2\sqrt{2H(\varepsilon_1)}\varepsilon_1 + \sum_{k=2}^{j} 2\sqrt{2H(\varepsilon_k)}\varepsilon_k = \sum_{k=1}^{j} 2\sqrt{2H(\varepsilon_k)}\varepsilon_k$$

$$\le 4\sqrt{2} \sum_{k=1}^{j} \int_{\varepsilon_{k+1}}^{\varepsilon_k} \sqrt{H(\varepsilon)}d\varepsilon \le 4\sqrt{2} \int_{0}^{\varepsilon_1} \sqrt{H(\varepsilon)}d\varepsilon = 4\sqrt{2}\mathscr{D}(\sigma/2).$$

It remains to pass from S_j to entire space T. This will be done in several stages.

1. If T is a finite set, then we have $T = S_j$ for some j and the theorem's assertion is already proved.
2. If $T = \{t_i\}_{i=1}^{\infty}$ is an infinite countable set, we apply the proved result to finite sets $T_n = \{t_i\}_{i=1}^{n}$ and take into account that

$$\mathbb{E} \sup_{t \in T_n} X(t) \nearrow \mathbb{E} \sup_{t \in T} X(t),$$

   and that entropy characteristics of the space (T_n, ρ) do not exceed those of (T, ρ).
3. If T is an arbitrary set, we can find a countable dense subset $T_{\#}$ (otherwise Dudley integral is infinite and there is nothing to prove). Apply the proved result to $T_{\#}$ and consider a version of X that satisfies $\sup_T X = \sup_{T_{\#}} X$. The assertion of theorem is satisfied for this version. □

Remark 10.1 According to (7.4), a median of random variable $\sup_T X$ is majorated by its expectation. Therefore, one can also use Dudley estimate for evaluation of median, which can be conveniently combined with Concentration Principle (6.9).

Exercise 10.1 (Pisier theorem [147]) Let $X(t), t \in T$, be a centered random process (not necessarily Gaussian), satisfying $\sigma^2 = \sup_{t \in T} \mathbb{E}X(t)^2 < \infty$. Prove that

$$\mathbb{E} \sup_{t \in T} X(t) \le 4 \int_{0}^{\sigma} \sqrt{N(\varepsilon)}d\varepsilon.$$

An important task of evaluation of *absolute* values of a process does not require new ideas due to the following result.

Proposition 10.2 *For any centered Gaussian process $X(t), t \in T$, it is true that*

$$\mathbb{E}\sup_{t\in T}|X(t)| \le 2\left(\mathbb{E}\sup_{t\in T}X(t) + \inf_{t\in T}(\mathbb{V}ar\,X(t))^{1/2}\right). \tag{10.4}$$

Proof (of Proposition 10.2). We use a standard notation $x_+ = \max\{x,0\}$, $x_- = \max\{-x,0\}$. It is true that $|x| = x_+ + x_-$, $(-x)_- = x_+$. Let us remark that

$$\mathbb{E}\sup_{t\in T}(X(t))_- = \mathbb{E}\sup_{t\in T}(-X(t))_- = \mathbb{E}\sup_{t\in T}(X(t))_+,$$

$$\mathbb{E}\sup_{t\in T}|X(t)| = \mathbb{E}\sup_{t\in T}((X(t))_+ + (X(t))_-)$$
$$\le \mathbb{E}\sup_{t\in T}(X(t))_+ + \mathbb{E}\sup_{t\in T}(X(t))_- = 2\mathbb{E}\sup_{t\in T}(X(t))_+. \tag{10.5}$$

On the other hand, we always have

$$\sup_{t\in T}(X(t))_+ \le \sup_{t\in T}X(t) + \inf_{t\in T}|X(t)|. \tag{10.6}$$

Indeed, if $\sup_{t\in T}X(t) \ge 0$, then

$$\sup_{t\in T}(X(t))_+ = \sup_{t\in T}X(t)$$

and inequality (10.6) holds, while if $\sup_{t\in T}X(t) < 0$, than both sides of (10.6) vanish. From (10.6), we find that

$$\mathbb{E}\sup_{t\in T}(X(t))_+ \le \mathbb{E}\sup_{t\in T}X(t) + \inf_{t\in T}\mathbb{E}|X(t)| \le \mathbb{E}\sup_{t\in T}X(t) + \inf_{t\in T}(\mathbb{V}ar\,X(t))^{1/2}.$$

By combining this inequality with (10.5), we obtain (10.4). $\qquad\square$

Remark 10.2 The second term is necessary in the bound (10.7): consider a singleton $T = \{t\}$. Then $\mathbb{E}\sup_T X = \mathbb{E}X(t) = 0$, and the first term alone does not provide a correct bound.

Proposition 10.3 (Sufficient condition for continuity) *Let $X(t), t \in T$, be a centered Gaussian process such that $\mathscr{D}(u) < \infty$, as $u > 0$. Then for any $t \in T$ the process X is a.s. continuous at t, i.e.*

$$\lim_{\rho(s,t)\to 0}X(s) = X(t). \tag{10.7}$$

Remark 10.3 Actually one can prove something more: if $\mathscr{D}(u) < \infty$, then with probability one (10.7) holds for all $t \in T$ simultaneously.

Proof (of Proposition 10.3). Fix $t \in T$ and let $T_n = \{s \in T : \rho(s,t) \le \frac{1}{n}\}$. Let $H_n(\cdot)$, $\mathscr{D}_n(\cdot)$ be the metric entropy, resp. Dudley integral for a process $\{X(s) - X(t), s \in T_n\}$. Let denote $S_n = \sup_{s\in T_n}(X(s) - X(t))$. It is clear that $H_n(\varepsilon) \le H(\varepsilon)$. Therefore, Theorem 10.1 yields a bound

$$\mathbb{E}S_n \leq 4\sqrt{2}\mathscr{D}_n\left(n^{-1}\right) \leq 4\sqrt{2}\mathscr{D}\left(n^{-1}\right).$$

The sequence of random variables S_n is non-negative and decreases to a limit S. By Fatou Lemma $\mathbb{E}S \leq \lim_n \mathbb{E}S_n = 0$. It follows that $S=0$ almost surely. □

The regularity of *stationary* processes should be mentioned separately. Here we have two nice results. The first of them shows that a stationary process is either continuous, or catastrophically discontinuous.

Theorem 10.2 (Belyaev alternative [17]) *Let $X(t)$, $t \in \mathbb{R}$, be a stationary Gaussian process. Then one of two incompatible assertions holds*

a) *Process X has continuous sample paths.*
b) *For any non-degenerate interval $T \subset \mathbb{R}$ we have $\sup_T X = \infty$ almost surely.*

The second result provides a necessary and sufficient condition for continuity of a stationary Gaussian process.

Theorem 10.3 (Fernique criterion [67]) *Let $X(t)$, $t \in \mathbb{R}$, be a stationary Gaussian process. Then X has continuous sample paths (with respect to the natural distance ρ) iff its Dudley integral is finite.*

For non-stationary processes Dudley integral does not provide continuity criterion: there exist continuous Gaussian processes with infinite Dudley integral. Moreover, it is known that necessary and sufficient conditions for continuity or boundedness can not be formulated in entropy terms. Stating and proving such conditions requires more subtle means, such as majorizing measures [67, 68, 107, 117, 166, 180] or generic chaining [173, 174]. We refer to [117] for detailed historical remarks, further bibliography of earlier works and recommend books [179, 180] for further reading on interesting applications of entropy bounds to non-Gaussian processes, real analysis, and ergodic theory.

10.3 Lower Bounds

The lower entropy bounds are based on the following result that we state here without proof.

Theorem 10.4 (Fernique–Sudakov comparison theorem [67, 162–164]) *Let $X(t)$, $Y(t)$ be two centered Gaussian processes defined on a common parametric set T. Assume that*

$$\mathbb{E}(X(t) - X(s))^2 \geq \mathbb{E}(Y(t) - Y(s))^2 \qquad \forall s, t \in T.$$

Then

$$\mathbb{E}\sup_{t \in T} X(t) \geq \mathbb{E}\sup_{t \in T} Y(t).$$

In other words, a process with larger increment variance has a larger expectation of supremum.

Comparison theorem is often conveniently combined with the following elementary bound that can be viewed as a conversion of Lemma 10.1. Notice, however, an independence assumption, that makes an important difference.

Lemma 10.2 *Let $(X_j)_{1\leq j\leq N}$ be independent centered Gaussian random variables and $\min_{j\leq N} \mathbb{E}X_j^2 \geq \sigma^2$. Let $c_* = 0.64$. Then*

$$\mathbb{E} \max_{1\leq j\leq N} X_j \geq c_*\sqrt{\ln N}\,\sigma, \tag{10.8}$$

Proof. Let $c < \sqrt{2}$. Let us show that

$$\lim_{N\to\infty} \mathbb{P}\left\{ \max_{1\leq j\leq N} X_j \leq c\sqrt{\ln N}\,\sigma \right\} = 0.$$

Indeed,

$$\mathbb{P}\left\{ \max_{1\leq j\leq N} X_j \leq c\sqrt{\ln N}\,\sigma \right\} = \prod_{j=1}^{N} \mathbb{P}\left\{ X_j \leq c\sqrt{\ln N}\,\sigma \right\}$$

$$\leq \left(1 - \widehat{\Phi}\left(c\sqrt{\ln N}\right)\right)^{N} \leq \exp\left\{ -N\widehat{\Phi}\left(c\sqrt{\ln N}\right) \right\},$$

where $\widehat{\Phi}$ is the tail of standard normal distribution. Apply an inequality

$$\widehat{\Phi}(u) \geq \frac{1}{\sqrt{2\pi}} \left(\frac{1}{u} - \frac{1}{u^3} \right) e^{-u^2/2}.$$

For large $u = c\sqrt{\ln N}$ we obtain

$$\widehat{\Phi}\left(c\sqrt{\ln N}\right) \geq \frac{N^{-c^2/2}}{3c\sqrt{\ln N}},$$

whence

$$\mathbb{P}\left\{ \max_{1\leq j\leq N} X_j \leq c\sqrt{\ln N}\,\sigma \right\} \leq \exp\left\{ -\frac{N^{1-c^2/2}}{3c\sqrt{\ln N}} \right\} \to 0, \quad \text{as } N \to \infty.$$

It follows that a median of random variable $\max_{1\leq j\leq N} X_j$ exceeds $c\sqrt{\ln N}\,\sigma$. For large N inequality (7.4) yields

$$\mathbb{E} \max_{1\leq j\leq N} X_j \geq c\sqrt{\ln N}\,\sigma.$$

Few initial values of N can only influence the constant in (10.8). Calculations show that $c_* = 0.64$ is an appropriate value for (10.8). $\qquad\square$

Theorem 10.5 *(Sudakov lower bound) For any $\varepsilon > 0$ and any centered Gaussian process $X(t), t \in T$, it is true that*

$$\mathbb{E} \sup_{t \in T} X(t) \geq \frac{c_*}{\sqrt{2}} \sqrt{\mathscr{C}(\varepsilon)} \varepsilon. \tag{10.9}$$

Proof. Let $\varepsilon > 0$ be fixed. Choose the points $\{t_1, \ldots, t_m\}$ in T such that $m = M(\varepsilon)$ and

$$\rho^2(t_i, t_j) = \mathbb{E}|X(t_i) - X(t_j)|^2 > \varepsilon^2, \quad i \neq j.$$

Let $Y_1, \ldots, Y_m$ be independent $N(0, \frac{\varepsilon^2}{2})$-distributed random variables. By applying Comparison Theorem 10.4 to the variables $(X(t_j))_{j \leq m}$ and $(Y_j)_{j \leq m}$, we obtain

$$\mathbb{E} \sup_{t \in T} X(t) \geq \mathbb{E} \sup_{j \leq m} X(t_j) \geq \mathbb{E} \sup_{j \leq m} Y_j.$$

Lemma 10.2 yields

$$\mathbb{E} \max_{1 \leq j \leq m} Y_j \geq c_* \sqrt{\ln m} \frac{\varepsilon}{\sqrt{2}} = \frac{c_*}{\sqrt{2}} \sqrt{\mathscr{C}(\varepsilon)} \varepsilon.$$

By combining two estimates, we arrive at (10.9). $\qquad \square$

Exercise 10.2 Let $X(t), t \in T$, be a centered Gaussian process such that the space (T, ρ) associated to X is compact. Let $c > 0$, $\varepsilon_0 > 0$ and $\delta \in (0, 2)$. Prove the following statements:

a) If $N(\varepsilon) \leq \exp\left(-c\varepsilon^{-(2-\delta)}\right)$, as $\varepsilon \in (0, \varepsilon_0)$, then the process X is almost surely continuous on (T, ρ).

b) If $N(\varepsilon) \geq \exp\left(-c\varepsilon^{-(2+\delta)}\right)$, as $\varepsilon \in (0, \varepsilon_0)$, then the process X is almost surely unbounded (hence, discontinuous) on (T, ρ).

Exercise 10.3 Let $X(t), t \in \mathbb{R}$, be a stationary Gaussian process with a spectral density

$$f(u) = |u|^{-1}(\ln |u|)^{-1-\beta}, \quad \beta > 0, |u| > 2.$$

Which range of parameter β corresponds to a continuous process X?

Exercise 10.4 Let $X(t), t \in [0, 1]$, be an α-fractional Brownian motion defined in Example 2.5. Find constants C_1 and C_2, depending on parameter α such that

$$C_1 \leq \mathbb{E} \sup_{t \in [0,1]} X(t) \leq C_2.$$

10.4 GB-Sets and GC-Sets

It is common to express the contents of this section in a geometrical language. Let $\mathcal{H}$ be a Hilbert space; we denote $(\cdot, \cdot)$ and $||\cdot||$ the corresponding scalar product and the norm. A centered Gaussian random function $X(h)$, $h \in \mathcal{H}$, is called *isonormal*, if

$$cov(X(h), X(h')) = (h, h'), \quad h, h' \in \mathcal{H}.$$

Note that the natural distance related to X is equal to that of Hilbert space,

$$\rho(h, h') = ||h - h'||, \quad h, h' \in \mathcal{H}. \tag{10.10}$$

Let $T \subset \mathcal{H}$. We call T a *GB-set* (resp. *GC-set*), if the restriction of an isonormal function $X(h)$, $h \in T$, possesses a bounded (resp. continuous) version.

All *GC*-sets are *GB*-sets, while all *GB*-sets are totally bounded (prove that both converse statements are wrong!). Therefore, the class of closed *GB*-sets, resp. *GC*-sets, is a subclass of compact sets. It is easy to see that both *GB* and *GC* classes are invariant with respect to the shifts and to the unitary rotations of $\mathcal{H}$.

Due to (10.10), we can forget about the process X and restate the results of current section regarding T as a metric subspace of $\mathcal{H}$ and using the notation $H_T(\cdot)$ and $\mathscr{C}_T(\cdot)$ for related metric entropy and metric capacity. In particular, by Theorem 10.5 a necessary condition

$$\sup_{\varepsilon > 0} \mathscr{C}_T(\varepsilon)\varepsilon^2 < \infty$$

is true for any *GB*-set T, while by Theorem 10.1 a sufficient condition

$$\int_0^\infty \sqrt{H_T(\varepsilon)} d\varepsilon < \infty$$

implies that T is a *GC*-set.

11 Small Deviations

11.1 Definitions and First Examples

In this section, a centered Gaussian $\mathscr{X}$-valued random vector is assumed to be given as a measurable mapping $X : (\Omega, \mathbb{P}) \mapsto \mathscr{X}$ taking values in a separable Banach space $(\mathscr{X}, ||\cdot||)$. As usual, we let P denote the distribution of X. Accordingly, H_P stands for the kernel of the measure P and $D := \{h \in H_P : |h|_{H_P} \le 1\}$ denotes the dispersion ellipsoid. Moreover, let $U := \{x \in \mathscr{X} : ||x|| \le 1\}$ denote the unit ball of $\mathscr{X}$. The *small deviation problem* (or *small ball problem*) suggests to explore

$$\mathbb{P}\left(\|X\| \le \varepsilon\right) = P\left(\varepsilon U\right), \quad \varepsilon \to 0.$$

A typical answer is

$$\mathbb{P}\left(\|X\| \le \varepsilon\right) \sim c_1 \varepsilon^a \exp\{-c_2 \varepsilon^{-b}\}, \quad \varepsilon \to 0. \tag{11.1}$$

In this case we call b *small deviation rate*, and c_2 *small deviation constant*. It is rarely possible to find a complete answer like (11.1) except for Markov processes including Wiener process and Brownian bridge. For example, for Wiener process [36, 146]

$$\mathbb{P}\left(\sup_{0 \le t \le 1} |W(t)| \le \varepsilon\right) \sim \frac{4}{\pi} \exp\{-\frac{\pi^2}{8}\varepsilon^{-2}\}, \quad \varepsilon \to 0, \tag{11.2}$$

and [35]

$$\mathbb{P}\left(\int_0^1 |W(t)|^2 dt \le \varepsilon^2\right) \sim \frac{4\varepsilon}{\sqrt{\pi}} \exp\{-\frac{1}{8}\varepsilon^{-2}\}, \quad \varepsilon \to 0, \tag{11.3}$$

while for Brownian bridge [92]

$$\mathbb{P}\left(\sup_{0 \le t \le 1} |W^0(t)| \le \varepsilon\right) \sim \frac{\sqrt{2\pi}}{\varepsilon} \exp\{-\frac{\pi^2}{8}\varepsilon^{-2}\}, \quad \varepsilon \to 0,$$

and [4]

$$\mathbb{P}\left(\int_0^1 |W^0(t)|^2 dt \le \varepsilon^2\right) \sim \frac{\sqrt{8}}{\sqrt{\pi}} \exp\{-\frac{1}{8}\varepsilon^{-2}\}, \quad \varepsilon \to 0.$$

One should notice that a small difference (rank one process) between Wiener process and Brownian bridge influences the power term in the asymptotics but not the exponential one.

11.2 Markov Case

We illustrate an approach to the study of small deviations for Markov processes by handling asymptotics (11.2). The self-similarity of Wiener process yields

$$\mathbb{P}\left(\sup_{0 \le t \le 1} |W(t)| \le \varepsilon\right) = \mathbb{P}\left(\sup_{0 \le t \le 1} |\varepsilon W(\varepsilon^{-2}t)| \le \varepsilon\right) = \mathbb{P}\left(\sup_{0 \le t \le \varepsilon^{-2}} |W(t)| \le 1\right)$$
$$= f(0, \varepsilon^{-2}),$$

where

$$f(x, T) := \mathbb{P}\left(\sup_{0 \le t \le T} |x + W(t)| \le 1 \right).$$

If $p(\cdot)$ denotes the distribution density of the standard normal law $N(0, 1)$, then Markov property of Wiener process and Taylor expansion actually yield, for small δ,

$$f(x, T) = o(\delta) + \int_{\frac{x-1}{\sqrt{\delta}}}^{\frac{x+1}{\sqrt{\delta}}} f(x + \sqrt{\delta} y, T - \delta) p(y) dy$$

$$\approx \int_{-\infty}^{\infty} \left(f(x, T) + f_x'(x, T)\sqrt{\delta} y \right.$$

$$\left. + \frac{1}{2} f_{xx}''(x, T)\delta y^2 - f_T'(x, T)\delta \right) p(y) dy$$

$$= f(x, T) + \frac{1}{2} f_{xx}''(x, T)\delta - f_T'(x, T)\delta.$$

After cancellation, we obtain the classical heat equation

$$\frac{1}{2} f_{xx}''(x, T) = f_T'(x, T)$$

while the boundary conditions are

$$f(1, T) = f(-1, T) = 0, \ f(x, 0) \equiv 1.$$

Looking at the solutions with separated variables for this equation, $f_k(x, T) = g_k(x)e^{-\lambda_k T}$, observe that the boundary conditions imply $g_k(\pm 1) = 0$. By plugging into the heat equation, we arrive at the ordinary differential equation

$$g_k''(x) + 2\lambda_k g_k(x) = 0, \quad |x| < 1,$$

$$g_k(\pm 1) = 0,$$

whence $g_k(x) = \cos(\pi(k + 1)/2)x)$ and $2\lambda_k = \pi^2(k + 1/2)^2$ for $k = 0, 1, 2, \ldots$. Therefore, the solution we search for should be written as a series

$$f(x, T) = \sum_{k=0}^{\infty} c_k \cos(\pi(k + 1/2)x) \exp\{-\pi^2(k + 1/2)^2 T/2\},$$

and the coefficients c_k are adjusted to the initial boundary condition $f(x, 0) \equiv 1$, i.e. (by taking into account that cosine functions are orthogonal)

$$c_k = \frac{\int_{-1}^{1} \cos(\pi(k + 1/2)x) dx}{\int_{-1}^{1} \cos^2(\pi(k + 1/2)x) dx} = \frac{2(-1)^k}{\pi(k + 1/2)}.$$

The first term of the series determines the solution asymptotics at $T \to \infty$, i.e.

$$f(x, T) \sim c_0 \cos(\pi x/2) \exp\{-\pi^2 T/8\}, \quad T \to \infty.$$

By plugging in $x = 0$, $T = \varepsilon^{-2}$, we arrive at (11.2).

The study of small deviations for additive norms like (11.3) can be performed similarly, but instead of equation for probabilities $f(x, T)$ one derives a heat equation for Laplace transform [30], e.g. for

$$\tilde{f}(x, T) := \mathbb{E} \exp\left\{-\int_0^T |x + W(t)|^2 dt\right\}.$$

With some loss of precision, the formulas (11.2) and (11.3) admit an extension to weighted L_q-norms. For example,

$$\ln \mathbb{P}\left(\sup_{0 \le t \le 1} |\rho(t) W(t)| \le \varepsilon\right) \sim -\frac{\pi^2}{8} \|\rho\|_{L_2[0,1]}^2 \varepsilon^{-2}, \quad \varepsilon \to 0,$$

if the function ρ^2 is Riemann integrable, and

$$\ln \mathbb{P}\left(\int_0^1 |\rho(t) W(t)|^q dt \le \varepsilon^q\right) \sim -c(q) \|\rho\|_{L_m[0,1]}^2 \varepsilon^{-2}, \quad \varepsilon \to 0,$$

whenever $1 \le q < \infty$, $m = \frac{2q}{q+2}$ and the function ρ^m is Riemann integrable [112, 113, 122].

11.3 Direct Entropy Method

In a particular case when $\mathcal{X}$ is a space of continuous functions and $\|\cdot\|$ is a supremum-norm, one can obtain comprehensive bounds for small deviations in terms of entropy characteristics from Sect. 10.

Let $X(t)$, $t \in T$, be a centered Gaussian random process, and $N(\varepsilon)$ denote its covering numbers. We denote, as before, $\sigma^2 := \sup_T \mathbb{E} X(t)^2$ and $\rho^2(s, t) := \mathbb{E}(X(s) - X(t))^2$.

Theorem 11.1 [9] *Assume that*

$$N(\varepsilon) \le \Psi(\varepsilon), \quad \forall \varepsilon > 0, \tag{11.4}$$

where the majorizing function Ψ is continuous, non-decreasing and satisfies regularity assumption

$$\Psi(\varepsilon/2) \le C\Psi(\varepsilon), \quad \forall \varepsilon > 0. \tag{11.5}$$

Then

$$\ln \mathbb{P} \left\{ \sup_{s,t \in T} |X(s) - X(t)| \le C_0 \varepsilon \right\} \ge -C_1 \widetilde{\Psi}(\varepsilon), \quad 0 < \varepsilon < \sigma/2, \qquad (11.6)$$

where C_0 is a numerical constant, $C_1 = C_1(C)$ and

$$\widetilde{\Psi}(\varepsilon) = \int_{\varepsilon}^{\sigma} \frac{\Psi(u)}{u} du, \quad 0 < \varepsilon < \sigma/2.$$

Remark 11.1 The regularity condition prohibits using exponential majorants but allows the polynomial and the logarithmic ones.

Remark 11.2 Formula (11.6) evaluates small deviations of somewhat unusual sample path norm $\sup_{s,t \in T} |X(s) - X(t)|$, called the *range* of the random process X. However, the bound for the classical sup-norm follows easily. By using weak correlation inequality (7.8), we find that for any $t \in T$ and $\delta \in (0, 1)$ it is true that

$$\mathbb{P} \left\{ \sup_{s \in T} |X(s)| \le \varepsilon \right\} \ge \mathbb{P} \left\{ \sup_{s \in T} |X(s) - X(t)| \le (1-\delta)\varepsilon; |X(t)| \le \delta\varepsilon \right\}$$

$$\ge \mathbb{P} \left\{ \sup_{s \in T} |X(s) - X(t)| \le (1-\delta)^2 \varepsilon \right\} \mathbb{P} \left\{ |X(t)| \le \frac{\delta\varepsilon}{K_\delta} \right\}.$$

The second factor is of order ε and almost newer affects the logarithmic asymptotics of small deviation, i.e. taking into account (11.6) and (11.5) we have

$$\ln \mathbb{P} \left\{ \sup_{s \in T} |X(s)| \le \varepsilon \right\} \ge -\widetilde{\Psi}(\varepsilon/C_0) \approx -\widetilde{\Psi}(\varepsilon). \qquad (11.7)$$

Remark 11.3 If we replace assumption (11.5) with a stronger bilateral assumption

$$C'\Psi(\varepsilon) \le \Psi(\varepsilon/2) \le C\Psi(\varepsilon)$$

with some $C' > 1$ (which excludes logarithmic majorants and actually means that Ψ is a kind of regularly varying function of negative index), it is easy to check that $\widetilde{\Psi}(\varepsilon) \le \frac{C' \ln 2}{C'-1} \Psi(\varepsilon)$, and we arrive at *Talagrand bound* [105, 169] for small deviation probabilities

$$\ln \mathbb{P} \left\{ \sup_{s,t \in T} |X(s) - X(t)| \le C_0 \varepsilon \right\} \ge -C_1' \Psi(\varepsilon), \quad 0 < \varepsilon < \sigma/2. \qquad (11.8)$$

For example, if $T = [0, 1]$ and $X = W^{(\alpha)}$ is an α-fractional Brownian motion, then $N(\varepsilon) \sim \varepsilon^{-2/\alpha}$ and Talagrand bound yields

$$\ln \mathbb{P} \left\{ \sup_{0 \le t \le 1} |W^{(\alpha)}(t)| \le \varepsilon \right\} \ge -c(\alpha)\varepsilon^{-2/\alpha}. \qquad (11.9)$$

However, the next example shows that the function $\widetilde{\Psi}$ sometimes can not be replaced with Ψ, as done in (11.8).

Example 11.1 Let $T = \mathbb{N}$ and let $X(s)$, $s \in \mathbb{N}$, be independent centered Gaussian variables with variances $\sigma_s^2 = e^{-2s}$. It is easy to check that $N(\varepsilon) \approx |\ln \varepsilon|$, i.e. a majorant $\Psi(\varepsilon) \asymp |\ln \varepsilon|$ applies, whereas $\widetilde{\Psi}(\varepsilon) \asymp |\ln \varepsilon|^2$ and the bound (11.7) yields

$$\ln \mathbb{P}\left\{ \sup_{s \in \mathbb{N}} |X(s)| \leq \varepsilon \right\} \geq -|\ln \varepsilon|^2.$$

It is not hard to check that the order of this bound is sharp, i.e.

$$\ln \mathbb{P}\left\{ \sup_{s \in \mathbb{N}} |X(s)| \leq \varepsilon \right\} \asymp -|\ln \varepsilon|^2. \tag{11.10}$$

Exercise 11.1 Prove the bound (11.10).

The proof of Theorem 11.1 will be based on the following evaluation.

Lemma 11.1 *Let $(\varepsilon_k)_{k \geq 0}$ be a decreasing sequence of positive numbers and assume that $\varepsilon_0 \geq \sigma$. Let $(b_k)_{k \geq 0}$ be a summable sequence of non-negative numbers and $b = \sum_{k \geq 0} b_k$. Then*

$$\mathbb{P}\left\{ \sup_{s,t \in T} |X(s) - X(t)| \leq 2b \right\} \geq \prod_{k=0}^{\infty} \mathbb{P}\{\varepsilon_k |\xi| \leq b_k\}^{N(\varepsilon_{k+1})}, \tag{11.11}$$

where ξ is a standard normal random variable.

Proof (of Lemma 11.1). We use the chaining method introduced in the proof of Theorem 10.1. For each ε_k choose a minimal covering of T by sets of diameter not exceeding ε_k. Let us choose a point in every covering element. Let S_k denote the set of $N(\varepsilon_k)$ chosen points. Let us fix some mappings $\pi_k : T \mapsto S_k$ satisfying

$$\rho(x, \pi_k(x)) \leq \varepsilon_k, \quad \forall x \in T.$$

Notice that for $k=0$ we have $N(\varepsilon_0) = 1$, i.e. the set S_0 consists of a single element. Therefore,

$$\sup_{s,t \in S_0} |X(s) - X(t)| = 0.$$

Furthermore, we have a bound

$$|X(s) - X(t)| \leq |X(s) - X(\pi_k(s))| + |X(\pi_k(s)) - X(\pi_k(t))| + |X(\pi_k(t)) - X(t)|,$$

that yields

$$\sup_{s,t\in S_{k+1}} |X(s) - X(t)| \le \sup_{s,t\in S_k} |X(s) - X(t)| + 2 \sup_{t\in S_{k+1}} |X(t) - X(\pi_k(t))|.$$

By induction,

$$\sup_{s,t\in S_{n+1}} |X(s) - X(t)| \le 2 \sum_{k=0}^{n} \sup_{t\in S_{k+1}} |X(t) - X(\pi_k(t))|.$$

We come to the key point of the proof, application of Khatri–Šidak inequality (7.7), which yields

$$\mathbb{P}\left\{ \sup_{s,t\in S_{n+1}} |X(s) - X(t)| \le 2b \right\} \ge \mathbb{P}\left\{ \bigcap_{k=0}^{n} \left\{ \sup_{t\in S_{k+1}} |X(t) - X(\pi_k(t))| \le b_k \right\} \right\}$$

$$= \mathbb{P}\left\{ \bigcap_{k=0}^{n} \bigcap_{t\in S_{k+1}} \{|X(t) - X(\pi_k(t))| \le b_k\} \right\}$$

$$\ge \prod_{k=0}^{n} \mathbb{P}\{\varepsilon_k|\xi| \le b_k\}^{N(\varepsilon_{k+1})}.$$

By replacing the finite product in the right-hand side with the infinite one, we get a bound independent of n. A passage from S_n to the set T repeats formal reasonings from the proof of Theorem 10.1. $\square$

Proof (of Theorem 11.1). Let $\varepsilon \in (0, \sigma/2)$. In order to use the bound (11.11), we have to construct appropriate sequences (ε_k) and (b_k). We provide different constructions for $\varepsilon_k \le \varepsilon$ and $\varepsilon_k \ge \varepsilon$. Let us fix an $r \in (\frac{1}{2}, 1)$. The value of r will only affect the constants emerging in the proof. We consider the zone $\varepsilon_k \le \varepsilon$ first. Let $\varepsilon_k = 2^{-k}\varepsilon$, $b_k = r^k\varepsilon$, $k = 0, 1, \ldots$. Clearly,

$$b := \sum_{k=0}^{\infty} b_k = \frac{\varepsilon}{1-r}$$

is an appropriate bound. By using (11.4) and iterating assumption (11.5), we obtain

$$N(\varepsilon_{k+1}) = N(2^{-k-1}\varepsilon) \le \Psi(2^{-k-1}\varepsilon) \le C^{k+1}\Psi(\varepsilon).$$

Since $2r > 1$, standard tail estimates of the normal law yield

$$\mathbb{P}\{\varepsilon_k|\xi| \le b_k\} = \mathbb{P}\{|\xi| \le (2r)^k\} = 1 - \mathbb{P}\{|\xi| \ge (2r)^k\}$$

$$\ge \exp\left\{-2\mathbb{P}\{|\xi| \ge (2r)^k\}\right\} \ge \exp\left\{-4\exp[-(2r)^{2k}/2]\right\}.$$

Thus,

$$\prod_{k=0}^{\infty} \mathbb{P}\{\varepsilon_k|\xi| \le b_k\}^{N(\varepsilon_{k+1})} \ge \prod_{k=0}^{\infty} \exp\{-4\exp[-(2r)^{2k}/2]C^{k+1}\Psi(\varepsilon)\}$$

$$= \exp\left\{-4\sum_{k=0}^{\infty} \exp[-(2r)^{2k}/2]C^{k+1}\Psi(\varepsilon)\right\}$$

$$:= \exp\{-c(r)\Psi(\varepsilon)\}.$$

By taking into account the monotonicity of function Ψ and assumption $\varepsilon < \sigma/2$, we have

$$\widetilde{\Psi}(\varepsilon) \ge \int_{\varepsilon}^{2\varepsilon} \frac{\Psi(u)}{u}du \ge \ln 2 \cdot \Psi(2\varepsilon) \ge \frac{\ln 2}{C}\Psi(\varepsilon), \qquad (11.12)$$

and the required estimate follows by

$$\prod_{k=0}^{\infty} \mathbb{P}\{\varepsilon_k|\xi| \le b_k\}^{N(\varepsilon_{k+1})} \ge \exp\left\{-\frac{c(r)C}{\ln 2}\widetilde{\Psi}(\varepsilon)\right\}.$$

Let us now consider the zone $\varepsilon_k \ge \varepsilon$ where we build a *finite* system of levels. Choose $n = n(\varepsilon)$ such that

$$r^n \Psi(\varepsilon) \le \Psi(\sigma) < r^{n-1}\Psi(\varepsilon).$$

Let $\varepsilon_0 = \sigma$, while choosing next ε_k so that

$$\Psi(\varepsilon_k) = r^{n-k}\Psi(\varepsilon), \quad 1 \le k \le n.$$

In particular, we have $\varepsilon_n = \varepsilon$. For any $k \ge 1$ it is true that $\Psi(\varepsilon_k) \le r^{-1}\Psi(\varepsilon_{k-1})$. Notice that

$$\ln\left(\frac{\varepsilon_{k-1}}{\varepsilon_k}\right)\Psi(\varepsilon_k) \le \int_{\varepsilon_k}^{\varepsilon_{k-1}} \frac{du}{u}r^{-1}\Psi(\varepsilon_{k-1}) \le r^{-1}\int_{\varepsilon_k}^{\varepsilon_{k-1}} \frac{\Psi(u)du}{u}. \qquad (11.13)$$

Finally, we define

$$b_k = r^{n-k}\varepsilon, \quad 0 \le k < n.$$

Since

$$\frac{b_k}{\varepsilon_k} = \frac{r^{n-k}\varepsilon}{\varepsilon_k} = r^{n-k}\frac{\varepsilon_n}{\varepsilon_k} \le 1,$$

we can use the simplest estimate

$$\mathbb{P}(\varepsilon_k|\xi| \le b_k) \ge c\frac{b_k}{\varepsilon_k} = c\frac{r^{n-k}\varepsilon_n}{\varepsilon_k},$$

where $c = (2/\pi)^{1/2}$. It follows that

$$\prod_{k=0}^{n-1} \mathbb{P}\left(\varepsilon_k |\xi| \le b_k\right)^{\Psi(\varepsilon_{k+1})} \ge \prod_{k=0}^{n-1} \left(c\frac{r^{n-k}\varepsilon_n}{\varepsilon_k}\right)^{\Psi(\varepsilon_{k+1})} := \Pi_1 \cdot \Pi_2,$$

where

$$\Pi_1 := \prod_{k=0}^{n-1} \left(cr^{n-k}\right)^{\Psi(\varepsilon_{k+1})}, \quad \Pi_2 := \prod_{k=0}^{n-1} \left(\frac{\varepsilon_n}{\varepsilon_k}\right)^{\Psi(\varepsilon_{k+1})}.$$

For Π_1 we easily obtain

$$|\ln \Pi_1| \le \sum_{k=0}^{n-1} \left(|\ln c| + |\ln r|(n-k)\right) r^{n-k-1} \Psi(\varepsilon) \le c(r)\Psi(\varepsilon),$$

the latter quantity being already evaluated in (11.12). For Π_2 summation by parts and taking (11.13) into consideration yield

$$|\ln \Pi_2| \le \sum_{k=0}^{n-1} \sum_{l=k+1}^{n} \ln\left(\frac{\varepsilon_{l-1}}{\varepsilon_l}\right) \Psi(\varepsilon_{k+1})$$

$$\le (1-r)^{-1} \sum_{l=1}^{n} \ln\left(\frac{\varepsilon_{l-1}}{\varepsilon_l}\right) \Psi(\varepsilon_l)$$

$$\le r^{-1} \int_{\varepsilon_n}^{\varepsilon_0} \frac{\Psi(u)\mathrm{d}u}{u} = r^{-1}\widetilde{\Psi}(\varepsilon).$$

It remains to merge two sequences ε_k, renumber the resulting sequence, and apply Lemma 11.1.   □

The estimates coming from the direct entropy method are simple to apply and often, although not always, efficient. They are, however missing generality in what concerns the norms to consider, they do not provide *upper* bounds for small deviation probabilities, and their results are just not sharp in the more complicated situations. This is why we pass now to a less elementary but by far more general *dual* entropy method.

11.4 Dual Entropy Method

Let X be a centered Gaussian vector taking values in a separable Banach space $(\mathscr{X}, \|\cdot\|)$. In the absence of Markov property, one can usually only pretend to find the rate of small deviations. Therefore, we will concentrate our efforts on the study of *small deviation function*

$$\phi(\varepsilon) := -\ln \mathbb{P}(\|X\| \le \varepsilon),$$

which satisfies the relation $\phi(\varepsilon) \asymp \varepsilon^{-b}$, as $\varepsilon \to 0$, for many random vectors X.

Following Kuelbs and Li, the authors of dual entropy method [98, 115], let us consider packing numbers $M_D(\varepsilon)$ of dispersion ellipsoid D with respect to the distance induced by the norm $\|\cdot\|$, and the corresponding metric capacity $\mathscr{C}_D(\varepsilon) = \ln M_D(\varepsilon)$. The word "dual" in subsection title stresses the fact that the nature of entropy characteristics we handle here is *different* from those considered in Sect. 10. The relation between two kinds of entropy is discussed below in Sect. 11.5.

We have the following upper bound for small deviation probability (which means a lower bound for small deviation function).

Proposition 11.11 *For any $r > 0$, $\lambda > 0$ it is true that*

$$\phi(r) \ge \mathscr{C}_D\left(\frac{2r}{\lambda}\right) - \frac{\lambda^2}{2}. \tag{11.14}$$

Proof. Let $U := \{x \in \mathscr{X} : \|x\| \le 1\}$ and let P denote the distribution of X in $\mathscr{X}$. Let $\varepsilon > 0$. Set $n = M_D(\varepsilon)$. Consider a point configuration $\{h_j, 1 \le j \le n\} \subset D$ such that $\|h_i - h_j\| > \varepsilon$ whenever $j \ne i$. Then all balls $h_j + \frac{\varepsilon}{2}U$ are disjoint and the same is true for dilated balls $\lambda(h_j + \frac{\varepsilon}{2}U) = \lambda h_j + \frac{\lambda \varepsilon}{2}U$. By using Borell inequality for shifted sets (5.3), we obtain

$$1 \ge \sum_{j=1}^{n} P\left(\lambda h_j + \frac{\lambda \varepsilon}{2}U\right)$$

$$\ge n \min_{1 \le j \le n} P\left(\frac{\lambda \varepsilon}{2}U\right) \exp\left\{-\frac{\lambda^2 |h_j|_{H_P}^2}{2}\right\}$$

$$\ge nP\left(\frac{\lambda \varepsilon}{2}U\right) \exp\left\{-\frac{\lambda^2}{2}\right\}.$$

Therefore,

$$P\left(\frac{\lambda \varepsilon}{2}U\right) \le \exp\left\{\frac{\lambda^2}{2}\right\} n^{-1}$$

and

$$\ln P\left(\frac{\lambda \varepsilon}{2}U\right) \le \frac{\lambda^2}{2} - \ln n.$$

Finally, we change notation by letting $r := \frac{\lambda \varepsilon}{2}$, then $\varepsilon = \frac{2r}{\lambda}$, and obtain

$$\ln P(rU) \le \frac{\lambda^2}{2} - \ln M_D\left(\frac{2r}{\lambda}\right),$$

which is equivalent to inequality (11.14).   □

Corollary 11.1 *Assume that for some $\beta \in (0, 2)$, $c > 0$ it is true that $\mathscr{C}_D(\varepsilon) \geq c\varepsilon^{-\beta}$ for small $\varepsilon > 0$. Then there exists $\tilde{c} > 0$ such that $\phi(r) \geq \tilde{c}\, r^{-\frac{2\beta}{2-\beta}}$ holds for all small $r > 0$.*

Proof (of Corollary). Fix $\delta > 0$. By plugging $\lambda := \delta r^{-\frac{\beta}{2-\beta}}$ in (11.14) we obtain

$$\ln P(rU) \leq \frac{\delta^2}{2r^{\frac{2\beta}{2-\beta}}} - \mathscr{C}_D\left(\frac{2r \cdot r^{\frac{\beta}{2-\beta}}}{\delta}\right),$$

$$\leq \frac{\delta^2}{2r^{\frac{2\beta}{2-\beta}}} - c\left(\frac{2r^{\frac{2}{2-\beta}}}{\delta}\right)^{-\beta}$$

$$= \left(\frac{\delta^2}{2} - c2^{-\beta}\delta^\beta\right)r^{\frac{-2\beta}{2-\beta}} := -\tilde{c}\,r^{\frac{-2\beta}{2-\beta}},$$

where $\tilde{c} > 0$ if δ is small enough.   □

Exercise 11.2 By using estimate (11.14), prove that $\mathscr{C}_D(\varepsilon) = o\left(\varepsilon^{-2}\right)$ holds for any centered Gaussian vector in a separable Banach space. Therefore the range $\beta \geq 2$ is senseless in the context of Corollary 11.1.

We turn now to the converse bounds. Let $N_D(\varepsilon)$ denote the covering numbers of dispersion ellipsoid D with respect to the distance induced by the norm $|| \cdot ||$ and let $H_D(\varepsilon) = \ln N_D(\varepsilon)$ denote the corresponding metric entropy. The lower bound for small deviation probabilities, i.e. the upper bound for small deviation function is as follows.

Proposition 11.2 *For any $r > 0$ it is true that*

$$\phi(2r) \leq \ln 2 + H_D\left(\frac{r}{\sqrt{2\Phi(r)}}\right). \tag{11.15}$$

Proof. Let $r, \lambda > 0$ and $n = N_D\left(\frac{r}{\lambda}\right)$. Let us cover ellipsoid D with a minimal number of sets with diameters not exceeding $\frac{\varepsilon}{\lambda}$; then inscribe each of these sets in a ball of the same radius. We obtain a covering

$$D \subset \bigcup_{j=1}^{n}\left\{h_j + \frac{r}{\lambda}U\right\}.$$

Multiplying by λ, we obtain

$$\lambda D \subset \bigcup_{j=1}^{n}\{\lambda h_j + rU\}.$$

It follows that

$$\lambda D + rU \subset \bigcup_{j=1}^{n} \{\lambda h_j + 2rU\}$$

and Anderson inequality yields

$$P(\lambda D + rU) \leq \sum_{j=1}^{n} P(\lambda h_j + 2rU) \leq nP(2rU).$$

On the other hand, from isoperimetric inequality it follows that

$$P(\lambda D + rU) \geq \Phi\left(\Phi^{-1}(P(rU)) + \lambda\right).$$

By comparing these bounds we obtain

$$\Phi\left(\Phi^{-1}(P(rU)) + \lambda\right) \leq nP(2rU).$$

In order to simplify the left-hand side, let $\lambda = \sqrt{2\phi(r)}$. Then

$$\Phi(-\lambda) \leq \exp\{-\lambda^2/2\} = \exp\{-\phi(r)\} = P(rU),$$

hence

$$-\lambda \leq \Phi^{-1}(P(rU)),$$

and we arrive at the estimate

$$nP(2rU) \geq \Phi(0) = \frac{1}{2}.$$

By taking logarithms, we have

$$\ln P(2rU) \geq -\ln 2 - \ln n,$$

which is equivalent to the required bound (11.15). □

Remark 11.4 Inequality (11.15) has an obvious drawback—an iterative nature. Indeed, the function $\phi(\cdot)$ shows up both in the left-hand and in the right-hand sides. However, for practically interesting examples the "degree" of $\phi(\cdot)$ in the right-hand side is less than that in the left-hand side. This observation explains the usefulness of (11.15). In particular, one can derive the converse estimates to the bounds stated above.

Corollary 11.2 *Assume that for some $\beta \in (0, 2)$, $c > 0$ we have $N_D(\varepsilon) \leq c\varepsilon^{-\beta}$ for small $\varepsilon > 0$. Then for some $\tilde{c} > 0$ the relation $\phi(r) \leq \tilde{c}r^{-\frac{2\beta}{2-\beta}}$ holds for small $r > 0$.*

Unfortunately, the known proof of this fact [115] based on (11.15) also involves the notions and results that are not considered in this course. It would be very interesting to find a short and self-contained proof for it.

Example 11.2 (Small deviations for Riemann–Liouville process [114]) Recall that α-Riemann–Liouville process was defined in Example 3.4. According to (4.5), its dispersion ellipsoid has the following form:

$$D = \left\{ h(t) = \frac{1}{\Gamma(\alpha)} \int_0^t (t-s)^{\alpha-1} \ell(s) ds, \, \|\ell\|_{L_2[0,1]} \le 1 \right\}.$$

The entropy characteristics of this function class in $\mathbb{C}[0,1]$ are studied quite well. It is known that $H_D(\varepsilon) \approx \varepsilon^{-1/\alpha}$. It follows that $\phi(r) \approx r^{-\frac{2}{2\alpha-1}}$. One can even show that a sharper statement $\phi(r) \sim Q(\alpha) r^{-\frac{2}{2\alpha-1}}$ is true for some constant $Q(\alpha)$.

Example 11.3 (Small deviations for fractional Brownian motion [114]) By comparing the integral representations of α-fractional Brownian motion with those of $\frac{\alpha+1}{2}$-Riemann–Liouville process, cf. (3.4) and (3.6), one can state that the difference of the two processes (up to normalizing constants), is a smooth process with relatively large small deviation probabilities. By using comparison estimates (7.9) and (7.10), one can show that the small deviation functions of two processes are equivalent. Taking into account the result of the previous example we see that small deviation function for α-fractional Brownian motion of satisfies

$$\phi(r) \sim Q\left(\frac{\alpha+1}{2}\right) \left(\frac{r}{c_\alpha \Gamma(\frac{\alpha+1}{2})} \right)^{-2/\alpha}.$$

where c_α is the constant from formula (3.5). This agrees with the one-sided bound (11.9) obtained by completely different method.

11.5 Duality of Metric Entropy

At first glance, it looks like the two entropies (that of the process parametric set with respect to its associated distance and that of dispersion ellipsoid of a Gaussian vector in a normed space) considered in previous subsections have nothing to do with each other. In fact, there is a deep connection between them coming from *duality conjecture* in linear operator theory. Let us forget for a while probability theory and small deviations and consider a problem in the language of linear operators. Let

$$V : (\mathscr{X}_1, \|\cdot\|_1) \to (\mathscr{X}_2, \|\cdot\|_2)$$

be a compact linear operator acting from one Banach space to another one. Let $B_1 = \{x \in \mathscr{X}_1 : \|x\|_1 \le 1\}$, $B_2 = \{x \in \mathscr{X}_2 : \|x\|_2 \le 1\}$ denote the corresponding unit balls. Compactness of V means that the set $V(B_1)$ is compact in $\mathscr{X}_2$. Therefore, we may consider covering numbers $N_{V(B_1)}(\varepsilon)$ of this set with respect to the distance of the space $\mathscr{X}_2$ and call them covering numbers of operator V,

$$N_V(\varepsilon) := N_{V(B_1)}(\varepsilon).$$

Function $N_V(\cdot)$ is a measure of complexity of operator V. Along with operator V, let us consider the dual operator

$$V^* : (\mathscr{X}_2^*, \|\cdot\|_{*,2}) \to (\mathscr{X}_1^*, \|\cdot\|_{*,1}).$$

The properties of operators V and V^* are tightly connected: their norms are equal and if V is compact, so is V^*. Therefore, a legitimate question arises about the connection between the covering numbers of operators V and V^*. In 1972, A. Pietsch stated *duality conjecture* for covering numbers which is still neither fully proved nor disproved. This conjecture asserts that there exist two numerical constants a and b such that for any linear operator V and any $\varepsilon > 0$ it is true that

$$b^{-1} \ln N_{V^*}(a\varepsilon) \le \ln N_V(\varepsilon) \le b \ln N_{V^*}(a^{-1}\varepsilon).$$

The essence of the statement is contained in the first inequality, since the second one follows by application of the first to V^*. Duality conjecture is proved for the case when at least one of the spaces we deal with is a Hilbert space [6].

Let us come back to a Gaussian vector $X \in \mathscr{X}$ we are interested in. Consider canonical embedding operators $I^* : \mathscr{X}^* \to \mathscr{X}_p^*$ and $I : \mathscr{X}_p^* \to \mathscr{X}$. The image of the unit ball under operator I coincides with dispersion ellipsoid. Therefore, the covering numbers $N_I(\varepsilon)$ are exactly those numbers on which the dual entropy approach to small deviations is based. Let us now write the norm of X as a supremum,

$$\|X\| = \sup_{f \in B^*} (f, X),$$

where B^* stands for the unit ball of the dual space $\mathscr{X}^*$. By applying the technique of direct entropy approach to $T = B^*$ and to the random process $\widetilde{X}(f) := (X, f)$ we have to study the entropy of the set B^* equipped with the distance

$$\rho(f,g)^2 = \mathbb{E}(\widetilde{X}(f) - \widetilde{X}(g))^2 = \mathbb{E}(f - g, X)^2 = \|I^* f - I^* g\|_{\mathscr{X}_p^*}^2.$$

Therefore, for the direct entropy approach we need exactly the covering numbers of operator I^*. We conclude that the covering numbers related to our approaches correspond to dual operators! Since the space $\mathscr{X}_p^*$ is a Hilbert space, the duality relations are true in the case we are interested in and, if necessary, we can replace one type of covering numbers with another one.

11.6 Hilbert Space

If $\mathscr{X}$ is a separable Hilbert space, the small deviation problem admits much sharper solution than in general case. Assuming that Karhunen–Loève expansion (2.1) is known, we can write the norm of a centered Gaussian vector as

$$||X||^2 = \sum_{j=1}^{\infty} \sigma_j^2 \xi_j^2, \tag{11.16}$$

(recall that ξ_j are independent standard normal variables), thus the small deviation problem reduces to a study of sums of independent variables. Formula (11.16) yields an explicit expression for Laplace transform,

$$\mathbb{E} \exp\{-\gamma ||X||^2\} = \prod_{j=1}^{\infty} (1 + 2\gamma \sigma_j^2)^{-1/2}.$$

Since the behavior of the norm distribution near zero (small deviations) is tightly related to the behavior of Laplace transform at large γ, we infer that everything is determined by the asymptotic behavior of the sequence (σ_j). For example, very accurate calculations yield the following result [88].

Proposition 11.3 *If $\alpha > 1$, $\theta \in \mathbb{R}$, and $\sigma_j^2 \sim C \frac{(\ln j)^{\theta}}{j^{\alpha}}$, then*

$$\phi(\varepsilon) \sim C^{\frac{1}{\alpha-1}} \left(\frac{\alpha-1}{2} \right)^{1-\frac{\theta}{\alpha-1}} \beta^{\frac{\alpha}{\alpha-1}} |\ln \varepsilon|^{\frac{\theta}{\alpha-1}} \varepsilon^{\frac{-2}{\alpha-1}}.$$

where $\beta := \frac{\pi}{\alpha \sin(\pi/\alpha)}$. Moreover, if the two-term asymptotics is known, i.e.

$$\sigma_j^2 = C \left(j + \delta + O\left(j^{-1} \right) \right)^{-\alpha}$$

then

$$P(||X|| \leq \varepsilon) \sim M \varepsilon^{\gamma} \exp \left\{ -C^{\frac{1}{\alpha-1}} \frac{\alpha-1}{2} \beta^{\frac{\alpha}{\alpha-1}} \varepsilon^{\frac{-2}{\alpha-1}} \right\},$$

where $\gamma = \frac{2-\alpha-2\delta\alpha}{2(\alpha-1)}$ and M depends on the entire sequence (σ_j).

However, this is only a first step to solution of small deviation problem, because a Gaussian vector is usually given not in the form of expansion (2.1).

Recall a typical situation with $\mathscr{X} = L_2(T, \mu)$ and a process given by covariance $K(s, t) = \mathbb{E}X(s)X(t)$. In this case, before applying Proposition 11.3, one should investigate the asymptotic behavior of unknown sequence (σ_j). Many new results and interesting examples appeared recently in this direction, see [16, 60, 69, 71–74, 88, 137–141].

Small deviation theory in Hilbert space admits two completely different natural extensions: to the sums of weighted i.i.d. variables [7, 31, 56, 119] and to L_p-spaces with arbitrary finite p [7, 61–64, 113, 122–124].

11.7 Other Results

In Corollaries 11.1 and 11.2 we considered criteria for power behavior of small deviation function. However, in certain cases the logarithmic factors also come into play. Therefore, the following extension [98, 115] is also useful.

Proposition 11.4 Let $\beta \in (0, 2)$, $\gamma \in \mathbb{R}$. Then relations $H_D(\varepsilon) \approx \varepsilon^{-\beta} |\ln \varepsilon|^{\gamma}$ and $\phi(r) \approx r^{-\frac{2\beta}{2-\beta}} |\ln r|^{\frac{2\gamma}{2-\beta}}$ are equivalent.

For example, the following result can be deduced via Proposition 11.4 from the entropy estimates of classes of smooth functions [15, 175]. Consider d-parametric Brownian sheet $X = W$ as a random element of the space $L_p[0, 1]^d$, $1 \le p < \infty$. Then $H_D(\varepsilon) \approx \varepsilon^{-1} |\ln \varepsilon|^{d-1}$ and $\phi(r) \approx r^{-2} |\ln r|^{2d-2}$. The small deviation behavior of the Brownian sheet with respect to the uniform norm (case $p = \infty$) is a difficult and challenging open problem. It is known that

$$
\phi(r) \begin{cases} \sim \frac{\pi^2}{8} r^{-2}, & d = 1, \\ \approx r^{-2} |\ln r|^3, & d = 2, [170], \\ \le r^{-2} |\ln r|^{2d-1} & d > 2, [55], \\ \ge r^{-2} |\ln r|^{2d-2+q}, & \text{for some } q = q(d) > 0, d > 2, [18]. \end{cases}
$$

Observe a logarithmic gap between the upper and lower bounds for $d > 2$.

For very smooth processes the small deviation function may even behave logarithmically. In such situations, another result is useful [8].

Proposition 11.5 Let $\gamma > 0$, $\zeta \in \mathbb{R}$. Then relations $H_D(\varepsilon) \approx |\ln \varepsilon|^{\gamma} \ln |\ln \varepsilon|^{\zeta}$ and $\phi(r) \approx |\ln r|^{\gamma} \ln |\ln \varepsilon|^{\zeta}$ are equivalent.

In Bayesian statistics one often considers a stationary Gaussian process with normal spectral density $f(u) = \exp(-u^2)$ for modelling (the logarithms of) random distribution densities. The behavior of the corresponding small deviation probabilities turns out to be crucial for investigation of convergence rates of Bayesian estimates [176]. Consider the mentioned process as an element of a family of processes X_{ν} with spectral densities

$$
f_{\nu}(u) = \begin{cases} \exp(-|u|^{\nu}), & 0 < \nu < \infty, \\ \mathbf{1}_{[0,1]}(u), & \nu = \infty. \end{cases}
$$

By using Proposition 11.5 jointly with the known estimates for the entropy of the classes of analytical functions [93], one can derive that for X_{ν} it is true that [8]

$$
\phi(r) \approx H_D(r) \approx \begin{cases} \frac{|\ln r|^2}{\ln |\ln r|}, & 1 < \nu \le \infty, \\ |\ln r|^{1+\frac{1}{\nu}}, & 0 < \nu \le 1. \end{cases}
$$

All examples mentioned above obey a general rule: *smoother the sample paths of a process are, larger are small deviation probabilities and smaller is the small deviation function.*

One can find a lot of additional information about small deviation theory in surveys [116, 120]. For an up to date bibliography on the topic, see [121].

12 Expansions of Gaussian Vectors

12.1 Problem Setting

On can state expansion problem for Gaussian vectors in two different ways.

Strong form. A Gaussian $\mathscr{X}$-valued random vector is given as a measurable mapping $X : (\Omega, \mathbb{P}) \mapsto \mathscr{X}$. Find an expansion

$$X(\omega) = \sum_{j=1}^{\infty} \xi_j(\omega) x_j$$

that holds $\mathbb{P}$-almost surely. Here ξ_j are independent Gaussian variables defined on $(\Omega, \mathbb{P})$ and (x_j) is a non-random sequence of vectors in $\mathscr{X}$.

Weak form. Given a Gaussian measure P on $\mathscr{X}$, construct a random vector X of the form

$$X(\omega) = \sum_{j=1}^{\infty} \xi_j(\omega) x_j$$

such that ξ_j are independent Gaussian variables, the series converges $\mathbb{P}$-almost surely and the distribution of X equals to P.

12.2 Series of Independent Random Vectors

For exposition simplicity, let us assume that $\mathscr{X}$ is equipped with a norm $\| \cdot \|$ and $(\mathscr{X}, \| \cdot \|)$ is a separable Banach space. Consider a series of partial sums

$$S_n = \sum_{j=1}^{n} X_j,$$

where $X_j \in \mathscr{X}$ are independent random vectors, defined on a common probability space $(\Omega, \mathbb{P})$. One can consider three types of convergence for S_n:

- distributions of S_n converge weakly to a distribution P on $\mathscr{X}$.

- There exists a limit vector $S \in \mathscr{X}$ such that $S_n \overset{\mathbb{P}}{\Rightarrow} S$, i.e.

$$\lim_{n \to \infty} \mathbb{P}\{||S_n - S|| > \varepsilon\} \qquad \forall \varepsilon > 0.$$

-
$$\lim_{n \to \infty} S_n = S \qquad \mathbb{P}\text{-almost surely.} \tag{12.1}$$

It is well known from general probability theory, that the third property yields the second, while the second yields the first one. For independent summands all three properties turn out to be equivalent.

Theorem 12.1 *If X_j are centered Gaussian vectors, the convergence of S_n in distribution implies (12.1).*

Remark 12.1 The Gaussian property in this theorem is not as important as independence. The statement remains true, e.g. for the series composed of the symmetrically distributed independent vectors.

Proof (of Theorem 12.1). We organize the proof in few steps, gradually enlarging the class of the spaces.

Step 1. $\mathscr{X} = \mathbb{R}^1$. Here X_j are just usual random variables. Let $\sigma_j^2 = \mathbb{V}ar X_j$. Then S_n has the distribution $N\left(0, \sum_{j=1}^n \sigma_j^2\right)$. Distribution convergence of S_n implies that $\sum_{j=1}^\infty \sigma_j^2 < \infty$. Since the series convergence for expectations and variances guarantees a.s.-convergence of a series of random variables (classical Kolmogorov–Khinchin "two-series theorem"), we see that (12.1) holds.

Step 2. $\mathscr{X} = \mathbb{R}^n$. The statement follows by applying the result of Step 1 to coordinates of vectors X_j. Recall that they are independent, symmetrically distributed, and for every coordinate the sum distributions converge to the distribution of this coordinate with respect to the limit law. On the other hand, it is obvious that the a.s. convergence of every coordinate of S_n yields a.s. convergence for vectors S_n.

Step 3. Let $\mathscr{X}$ be an arbitrary finite-dimensional space. The statement follows from the fact proved on the previous step by considering a linear isomorphism between $\mathscr{X}$ and the corresponding $\mathbb{R}^n$ such that this isomorphism and its inverse are bounded operators.

Step 4. Let $\mathscr{X}$ be an arbitrary separable Banach space. Let us first reformulate the notion of almost sure convergence in terms close to convergence in probability. We say that a sequence of random elements S_n of a normed space $\mathscr{X}$ is *fundamental in probability*, if

$$\lim_{n \to \infty} \mathbb{P}\left\{ \sup_{n_1, n_2 \geq n} ||S_{n_1} - S_{n_2}|| \geq \varepsilon \right\} = 0, \quad \forall \varepsilon > 0.$$

Let us show that if a sequence S_n is fundamental in probability and the space $\mathscr{X}$ is complete, then there exists a limiting random element $S \in \mathscr{X}$, such that $\lim_n S_n = S$ almost surely. Indeed, let

$$M_n = \sup_{n_1, n_2 \geq n} \|S_{n_1} - S_{n_2}\|.$$

Since M_n is a decreasing sequence, there exists a limit $M = \lim_n M_n$. For any $n \geq 1$, $\varepsilon > 0$ we have $\{M \geq \varepsilon\} \subset \{M_n \geq \varepsilon\}$. It follows that

$$\mathbb{P}\{M \geq \varepsilon\} \leq \lim_{n \to \infty} \mathbb{P}\{M_n \geq \varepsilon\} = 0.$$

Since ε is arbitrary, we have $M=0$ almost surely. Therefore, S_n is a Cauchy sequence, and the limit S exists.

Let us come back to our series. Assume that the sequence of partial sums converges in distribution. We will show that it is fundamental in probability. Let P be the limit distribution for S_n. Let us fix small numbers $\varepsilon, \delta > 0$. Since any finite measure in a separable Banach space is tight, there exists a compact set $K \subset \mathcal{X}$ such that $P(K) \geq 1 - \delta$. Let us choose a finite ε-net $x_1, \ldots, x_m$ in K, and let $\tilde{\mathcal{X}}$ be the linear span of this net. As usual, the distance between a point and a set is defined by the formula

$$\rho(x, A) = \inf_{y \in A} \|x - y\|, \quad x \in \mathcal{X}, A \subset \mathcal{X}.$$

Then

$$\sup_{x \in K} \rho(x, \tilde{\mathcal{X}}) \leq \sup_{x \in K} \inf_{1 \leq j \leq m} \|x - x_j\| \leq \varepsilon.$$

Therefore, convergence of distributions implies

$$\limsup_{n \to \infty} \mathbb{P}\{\rho(S_n, \tilde{\mathcal{X}}) \geq 2\varepsilon\} \leq P\{x : \rho(x, \tilde{\mathcal{X}}) > 2\varepsilon\}$$

$$\leq P(\mathcal{X} \backslash K) \leq \delta.$$

Therefore, for sufficiently large n it is true that

$$\mathbb{P}\{\rho(S_n, \tilde{\mathcal{X}}) \geq 2\varepsilon\} \leq 2\delta.$$

It is known from functional analysis that there exists a linear projector $L : \mathcal{X} \mapsto \tilde{\mathcal{X}}$ such that $\|L\| \leq 2$. Write

$$S_n = (S_n - LS_n) + LS_n := S_n' + \tilde{S}_n.$$

We derive now some estimates that enable to reduce the problem to the finite-dimensional case. Find a random element $y_n \in \tilde{\mathcal{X}}$ where the minimum is attained,

$$\|S_n - y_n\| = \rho(S_n, \tilde{\mathcal{X}}).$$

Then

$$\|S_n'\| = \|S_n - LS_n\| \le \|S_n - y_n\| + \|y_n - Ly_n\| + \|Ly_n - LS_n\|$$
$$\le (1 + \|L\|)\,\|S_n - y_n\| \le 3\rho(S_n, \tilde{\mathscr{X}}).$$

We use the following lemma for getting a uniform bound.

Lemma 12.1 (Lévy inequality) *If the random vectors X_j are independent and symmetrically distributed, then for any $r > 0$ their sums satisfy the bound*

$$\mathbb{P}\left\{\max_{k \le n} \|S_k\| \ge r\right\} \le 2\mathbb{P}\left\{\|S_n\| \ge r\right\}. \tag{12.2}$$

By applying (12.2) to the sums S_n', we obtain

$$\mathbb{P}\left\{\max_{k \le n} \|S_k'\| \ge 6\varepsilon\right\} \le 2\mathbb{P}\left\{\|S_n'\| \ge 6\varepsilon\right\} \le 2\mathbb{P}\left\{\rho(S_n, \tilde{\mathscr{X}}) \ge 2\varepsilon\right\} \le 2\delta.$$

Hence,

$$\mathbb{P}\left\{\max_{k} \|S_k'\| \ge 6\varepsilon\right\} = \lim_{n \to \infty} \mathbb{P}\left\{\max_{k \le n} \|S_k'\| \ge 6\varepsilon\right\} \le 2\delta.$$

On the other hand, the finite-dimensional sequence $\tilde{S}_n$ converges in distribution, being a linear projection of convergent sequence S_n. By using the result of Step 3, the sums $\tilde{S}_n$ are fundamental in probability. Therefore, for large n we obtain

$$\mathbb{P}\left\{\sup_{n_1, n_2 \ge n} \|S_{n_1} - S_{n_2}\| \ge 13\varepsilon\right\}$$
$$\le \mathbb{P}\left\{\sup_{n_1, n_2 \ge n} \|S_{n_1}' - S_{n_2}'\| \ge 12\varepsilon\right\} + \mathbb{P}\left\{\sup_{n_1, n_2 \ge n} \|\tilde{S}_{n_1} - \tilde{S}_{n_2}\| \ge \varepsilon\right\}$$
$$\le \mathbb{P}\left\{\sup_{k} \|S_k'\| \ge 6\varepsilon\right\} + \delta \le 3\delta.$$

Since δ was chosen arbitrarily, we have checked that the sequence S_n is fundamental in probability, and the proof is complete. □

12.3 Construction of a Vector with Given Distribution

Let a centered Gaussian distribution $P = N(0, K)$ be given and let H_P be the correspondent kernel. Let (ξ_j) be a sequence of independent $N(0, 1)$-distributed random variables and (z_j) an orthonormal base in $\mathscr{X}_P^*$. Then $(h_j) = (Iz_j)$ is an orthonormal base in H_P (Here I is the canonical isomorphism of spaces $\mathscr{X}_P^*$ and H_P, cf. Sect. 4). We show that the series

$$X(\omega) = \sum_{j=1}^{\infty} \xi_j(\omega)h_j$$

converges and that the distribution of series X coincides with P, [153]. According to Theorem 12.1, it is enough to check that partial sums $S_n = \sum_{j=1}^{n} \xi_j(\omega) h_j$ converge in distribution to P. We check the convergence of characteristic functions. Indeed, for $f \in \mathscr{X}^*$ we have

$$(f, S_n) = \sum_{j=1}^{n} \xi_j(\omega)(f, h_j) = \sum_{j=1}^{n} \xi_j(\omega)(I^* f, z_j). \qquad (12.3)$$

$$\mathbb{E} e^{i(f, S_n)} = \mathbb{E} \exp \left\{ i \sum_{j=1}^{n} \xi_j(\omega)(I^* f, z_j) \right\} = \exp \left\{ -\frac{1}{2} \sum_{j=1}^{n} (I^* f, z_j)^2 \right\}$$

$$\to \exp \left\{ -\frac{1}{2} \sum_{j=1}^{\infty} (I^* f, z_j)^2 \right\} = \exp \left\{ -\frac{1}{2} \|I^* f\|_2^2 \right\} = \int_{\mathscr{X}} e^{i(f, x)} P(\mathrm{d}x).$$

$\square$

12.4 Expansion of a Given Vector

We will solve now the strong form of expansion problem. Let a random vector X with distribution P be given. We will use the previous construction but specify the choice of random variables (ξ_j) by letting $\xi_j = z_j(X)$. Recall that measurable linear functionals z_j form a base in $\mathscr{X}_P^*$, thus

$$\mathbb{E} z_i(X) z_j(X) = \int_{\mathscr{X}} z_i z_j \mathrm{d}P = \begin{cases} 0, & i \neq j, \\ 1, & i = j. \end{cases}$$

We already know that the sum

$$Y = \sum_{j=1}^{\infty} z_j(X) h_j$$

is well defined $\mathbb{P}$-almost surely. It remains to prove that $Y=X$ with probability one. For any functional $f \in \mathscr{X}^*$ by (12.3) we have

$$(f, Y) = \lim_{n \to \infty} \sum_{j=1}^{n} z_j(X)(I^* f, z_j)$$

$\mathbb{P}$-almost surely. On the other hand, in the space $\mathscr{X}_P^* \subset L_2(\mathscr{X}, P)$ we have

$$\lim_{n\to\infty} \sum_{j=1}^{n} z_j(\cdot)(I^* f, z_j) = (I^* f)(\cdot),$$

i.e.

$$\mathbb{E}\left|\sum_{j=1}^{n} z_j(X)(I^* f, z_j) - (f, X)\right|^2 = 0.$$

It follows that $(f, X) = (f, Y)$ almost surely. Hence, for any $f \in \mathcal{X}^*$ we have $\mathbb{E} e^{i(f,X-Y)} = 1$ and $X = Y$ with probability one.   □

12.5 Examples: Expansions of Wiener Process

Consider some expansion examples by taking $\mathbb{C}[0, 1]$ as a space and a Wiener process W as a Gaussian vector. Recall that one can obtain a base in the kernel H_P of the Wiener measure P by integration of a base in $L_2[0, 1]$.

Example 12.1 (*Cosine base*) Let us consider a base in $L_2[0, 1]$ given by

$$\begin{cases} \varphi_0(s) = 1, \\ \varphi_j(s) = \sqrt{2}\cos(\pi j s), & j \geq 1. \end{cases}$$

Integration yields a base in the kernel

$$\begin{cases} h_0(t) = t, \\ h_j(t) = \sqrt{2}\frac{\sin(\pi j t)}{\pi j}, & j \geq 1. \end{cases}$$

We arrive at the expansion

$$W(t) = \xi_0 t + \sqrt{2}\sum_{j=1}^{\infty} \xi_j \frac{\sin(\pi j t)}{\pi j}.$$

Notice that $W(1) = \xi_0$ and use the representation (3.3) for Brownian bridge. We see that once the first (linear) term is dropped, the remaining sum presents an expansion for Brownian bridge.

Example 12.2 (*Sine base*) Take a base in $L_2[0, 1]$ given by

$$\varphi_j(s) = \sqrt{2}\sin(\pi j s), \quad j \geq 1.$$

Integration yields a base in the kernel

$$h_j(t) = \sqrt{2}\frac{1 - \cos(\pi j t)}{\pi j}, \quad j \geq 1.$$

We arrive at the expansion

$$W(t) = \sqrt{2} \sum_{j=1}^{\infty} \xi_j \frac{1 - \cos(\pi j t)}{\pi j}.$$

This expansion showed up in pioneer works of N. Wiener.

Example 12.3 (*Karhunen–Loève expansion*) Take a base in $L_2[0, 1]$ given by

$$\varphi_j(s) = \sqrt{2}\cos(\pi(j - 1/2)s), \quad j \geq 1.$$

Integration yields a base in the kernel

$$h_j(t) = \sqrt{2}\frac{\sin(\pi(j - \frac{1}{2})t)}{\pi(j - \frac{1}{2})}, \quad j \geq 1.$$

We arrive at the expansion

$$W(t) = \sqrt{2} \sum_{j=1}^{\infty} \xi_j \frac{\sin(\pi(j - \frac{1}{2})t)}{\pi(j - \frac{1}{2})}.$$

The functions h_j are also orthogonal in $L_2[0, 1]$ which is a particular advantage of this expansion. If one considers W as a random element of $L_2[0, 1]$ instead of $\mathbb{C}[0, 1]$, then (h_j) turns out to be the orthogonal eigenfunction system of the corresponding covariance operator.

Example 12.4 (*Paley–Wiener expansion,* [142]) Take a base in $L_2[0, 1]$ given by

$$\begin{cases} \varphi_0(s) = 1, \\ \varphi_{2j}(s) = \sqrt{2}\cos(2\pi j s), \quad j \geq 1, \\ \varphi_{2j-1}(s) = \sqrt{2}\sin(2\pi j s), \quad j \geq 1. \end{cases}$$

Integration yields a base in the kernel

$$\begin{cases} h_0(t) = t, \\ h_{2j}(t) = \frac{\sin(2\pi j t)}{\sqrt{2}\pi j}, \quad j \geq 1, \\ h_{2j-1}(t) = \frac{(1 - \cos(2\pi j t))}{\sqrt{2}\pi j}, \quad j \geq 1. \end{cases}$$

We arrive at the expansion

$$W(t) = \xi_0 t + \sum_{j=1}^{\infty} \xi_{2j} \frac{\sin(2\pi j t)}{\sqrt{2}\pi j} + \sum_{j=1}^{\infty} \xi_{2j-1} \frac{(1 - \cos(2\pi j t))}{\sqrt{2}\pi j}.$$

Example 1.12.5 (*Haar–Schauder expansion*) Let denote

$$\Psi(s) = \begin{cases} 1, & 0 \le s \le 1/2, \\ -1, & 1/2 < s \le 1, \\ 0, & s < 0 \text{ or } s > 1, \end{cases}$$

and

$$h(t) = \int_0^t \Psi(s)ds = \begin{cases} t, & 0 \le t \le \frac{1}{2}, \\ 1 - t, & \frac{1}{2} \le t \le 1, \\ 0, & t < 0 \text{ or } t > 1. \end{cases}$$

Consider the Haar base in $L_2[0, 1]$ given by

$$\begin{cases} \varphi_0(s) = 1, \\ \varphi_{j,k}(s) = 2^{j/2}\Psi\left(2^j\left(t - \frac{k}{2^j}\right)\right), & j \ge 0, \quad 0 \le k \le 2^j - 1. \end{cases}$$

Integration yields a base in the kernel

$$\begin{cases} h_0(t) = t, \\ h_{j,k}(t) = 2^{-j/2}h\left(2^j\left(t - \frac{k}{2^j}\right)\right), & j \ge 0, \quad 0 \le k \le 2^j - 1. \end{cases}$$

"Triangular functions" $h_{j,k}$ are called *Schauder functions*. We arrive at the expansion

$$W(t) = \xi_0 t + \sum_{j=0}^{\infty} \sum_{k=0}^{2^j-1} \xi_{j,k} h_{j,k}(t). \tag{12.4}$$

It is interesting that analogous representation for W on the half-line $[0, \infty)$ looks more homogeneously:

$$W(t) = \sum_{j=-\infty}^{\infty} \sum_{k=-\infty}^{\infty} \xi_{j,k} h_{j,k}(t).$$

All terms with negative j are linear functions on $[0,1]$. By summing them up we obtain the first term from (12.4).

The representation (12.4) is often called *Lévy construction* for Wiener process, [110]. Similar expansions emerge from other wavelet bases in $L_2[0, 1]$.

Representations similar to Lévy construction (using Schauder base) are available for more or less arbitrary Gaussian processes, see [39]. They proved to be very useful for studies of processes' properties, in particular, small deviations [159, 160].

Example 12.6 (*Expansion of complex Wiener process*, [142]) Consider a complex Wiener process $W(t) = W_1(t) + i W_2(t)$, where W_1, W_2 are independent real Wiener processes. Then W is a Gaussian random vector in the space of continuous complex-valued functions $\mathbb{C}_\mathbf{C}([0, 1])$.

Take a base in $L_{2,\mathbf{C}}[0, 1]$ given by

$$\varphi_j(s) = \exp\{2\pi i j s\}, \quad j \in \mathbb{Z}.$$

Integration yields a base in the kernel

$$h_j(s) = \frac{\exp\{2\pi i j t\} - 1}{2\pi i j}, \quad j \in \mathbb{Z}.$$

We arrive at the expansion

$$W(t) = \sum_{j=-\infty}^{\infty} \xi_j \frac{\exp\{2\pi i j t\} - 1}{2\pi i j}, \tag{12.5}$$

where (ξ_j) are independent *complex* standard Gaussian variables.

Example 12.7 (*Expansion of complex fractional Brownian motion*) The previous example admits an extension to the case of complex α-*fractional* Brownian motion, i.e. the process $W^{(\alpha)}(t) = W_1^{(\alpha)}(t) + i W_2^{(\alpha)}(t)$, where $W_1^{(\alpha)}$, $W_2^{(\alpha)}$ are independent real fBm (cf. Example 2.5). Similarly to (12.5), Dzhaparidze and van Zanten [57] found an expansion

$$W^{(\alpha)}(t) = \sum_{j=-\infty}^{\infty} \sigma_j \xi_j \frac{\exp\{2i\omega_j t\} - 1}{2i\omega_j}, \tag{12.6}$$

where (ξ_j) are again independent *complex* standard Gaussian variables, ω_j are the real zeros of Bessel function $J_{1-\alpha/2}(\cdot)$, and the coefficient variances are given by

$$\sigma_j^2 = \left[(2 - \alpha) \Gamma(1 - \alpha/2)^2 (\omega_j/2)^\alpha J_{-\alpha/2}(\omega_j) V \right]^{-1},$$

$$V = \frac{\Gamma(\frac{3-\alpha}{2})}{\alpha \Gamma(\frac{\alpha+1}{2}) \Gamma(3 - \alpha)}.$$

Let us note that for $\alpha = 1$ it is true that $J_{1/2}(z) = (2\pi/z)^{1/2} \sin z$. Therefore, $\omega_j = \pi j$ and (12.6) boils down to (12.5).

One can find variants of expansion (12.6) for other Gaussian processes and random fields in [57, 58, 133].

Exercise 12.1 Consider a Brownian sheet (Wiener–Chentsov field) W defined in Example 2.7 as a random element of the space $\mathbb{C}([0, 1]^2)$. Let H be the kernel of W. Construct an orthonormal base in H and build a series expansion of W as it was done for Wiener process.

12.6 Linear Operators, Associated Gaussian Vectors, and Their Expansions

Let $\mathscr{X}$ be a linear space satisfying our "usual assumptions", let $\mathscr{H}$ be a Hilbert space and consider a linear operator $J : \mathscr{H} \mapsto \mathscr{X}$. Taking any orthonormal base (e_j) in $\mathscr{H}$ we define a formal series

$$X = \sum_{j=1}^{\infty} \xi_j J e_j, \tag{12.7}$$

where (ξ_j) is a sequence of independent $N(0,1)$-distributed random variables.

One can show that the series (12.7) converges a.s. or diverges a.s., and the fact of convergence does not depend on the choice of the base (e_j). In the sequel we assume that the series (12.7) converges. Then X is a centered Gaussian vector with covariance operator $K = JJ^*$. Therefore, the distribution $P = N(0, JJ^*)$ of X does not depend of the base choice in $\mathscr{H}$. Moreover, by Factorization Theorem the kernel H_P of measure P coincides with $J(\mathscr{H})$. We say that a Gaussian vector X and measure P are *associated* to operator J.

If $(\mathscr{X}, || \cdot ||)$ is a normed space, we can define some important compactness characteristics of operator J in terms of vector X. The quantity

$$||J||_{\ell} := \left(\mathbb{E}||X||^2 \right)^{1/2}$$

is called ℓ-*norm* of operator J, cf. [126, 148]. The choice of the second moment for the norm definition is unimportant, since all moments of the norm of Gaussian vector are equivalent according to (8.10).

Next, *stochastic approximation numbers* $\ell_n(J)$ that characterize the quality of possible approximation of operator J by finite rank operators are defined as follows [115, 148]:

$$\ell_n(J) = \inf \left\{ ||J - F||_{\ell}; \ F : \mathscr{H} \mapsto \mathscr{X}, rank(F) < n \right\}.$$

One can show that $\ell_n(J)$ also characterize the quality of average approximation of associated random vector by Gaussian random vectors of finite rank:

$$\ell_n(J)^2 = \inf_{\substack{x_1,\dots,x_{n-1} \\ \xi_1,\dots,\xi_{n-1}}} \left\{ \mathbb{E} \left\| X - \sum_{j=1}^{n-1} \xi_j x_j \right\|^2 \right\}.$$

Notice that the convergence rate of the series (12.7) may depend on the base (e_j). However, the bases with good convergence rate do exist [100]:

Exercise 12.2 Let $\alpha > 0$, $\beta \in \mathbb{R}$. Prove that there exists a constant $C_{\alpha,\beta}$ such that for any operator J satisfying

$$\ell_n(J)^2 \le n^{-\alpha}(1 + \ln n)^\beta, \quad n \ge 1,$$

and any Gaussian vector X associated to J there exists a base (e_j) such that

$$\mathbb{E} \left\| X - \sum_{j=1}^{n-1} \xi_j J e_j \right\|^2 \le C_{\alpha,\beta} n^{-\alpha}(1 + \ln n)^\beta, \quad n \ge 1.$$

13 Quantization of Gaussian Vectors

13.1 Problem Setting

Quantization (or *discretization*) problem for random vectors comes from information processing. Imagine a signal (a fragment of a picture, of a soundtrack etc) that should be transmitted through a channel. The set of possible signals is a metric space $(\mathcal{X}, \rho)$. One of possible ideas for transmission algorithms is based on a use of "dictionaries". A *dictionary* is a finite subset $Y = \{y_j\}_{1 \le j \le n}$ of $\mathcal{X}$. Both the sender and receiver possess a copy of a dictionary. When a signal $x \in \mathcal{X}$ should be transmitted, the sender identifies a dictionary element y_j closest to x and sends the *number j* of approximating element through the channel. The receiver reconstructs the value y_j by using the dictionary. Clearly, the number transmission is performed much faster than the transmission of the signal itself. For this acceleration, we pay by making an error, since the output is an approximation y_j instead of the true signal x.

We will apply a Bayesian approach to the analysis of this transmission procedure. Assume that there is a probability measure P on $\mathcal{X}$ characterizing the probability distribution of a signal that eventually will be transmitted. Let X be a random element of $\mathcal{X}$ having distribution P. Then the average quantization error related to a dictionary Y is defined by

$$d(Y, p) = \left(\mathbb{E} \min_{1 \le j \le n} \rho(X, y_j)^p \right)^{1/p}.$$

Usually, the *high resolution* quantization, i.e. the procedure's asymptotic behavior is studied, as the dictionary size tends to infinity.

The problem is how to construct a reasonable dictionary of large size in a complicated signal space? One possibility to do it is to consider a *random* dictionary composed of random independent elements Y_j having the same distribution P as the prior signal distribution. We will provide now a quantitative study of this scheme. In order to work with polynomial functions, we make an exponential variable change $n = e^\lambda$. Now the quantization error reads as

$$D(\lambda, p) = \left(\mathbb{E} \min_{1 \le j \le e^\lambda} \rho(X, Y_j)^p \right)^{1/p}.$$

Following an extensive literature, we assume that $(\mathcal{X}, \|\cdot\|)$ is a normed space, $\rho(x, y) = \|x - y\|$, and the common distribution P of all random vectors is a centered Gaussian distribution.

13.2 Quantization and Small Deviations

Under these assumptions, as discovered in [49], quantization error of a Gaussian vector is tightly related to its small deviations' behavior, [49, 65]. Here is a clear demonstration of this link.

Theorem 13.1 *Let* $M_\lambda = \min_{j \le e^\lambda} \|Y_j - X\|$. *Then*

$$\lim_{\lambda \to \infty} \mathbb{P}\left(M_\lambda \ge 2\phi^{-1}(\lambda/6)\right) = 0.$$

Under mild additional assumptions, one can also evaluate more traditional moment characteristics $D(\lambda, p) = (\mathbb{E}M_\lambda^p)^{1/p}$.

Theorem 13.2 *Assume that the small deviation function of a Gaussian vector X satisfies regularity condition: for some* $c > 0$ *it is true that*

$$\phi(cr) \ge 2\phi(r), \quad r < r_0. \tag{13.1}$$

Then for any $p > 0$ *we have*

$$\limsup_{\lambda \to \infty} \frac{D(\lambda, p)}{\phi^{-1}(\lambda/2)} \le 2.$$

Let us remark, concerning the latter theorem, that the value of p is not important due to concentration property of Gaussian distribution.

Both theorems show that if small deviation probabilities are not too small, i.e. the functions ϕ, ϕ^{-1} admit upper bounds, then we can evaluate quantization error.

Proof (of Theorem 13.1). For $x \in \mathcal{X}$, $r > 0$ let

$$v(x, r) = \inf\{|h|_{H_P}, \|h - x\| \le r\}.$$

Then

$$\mathbb{P}(\|Y - x\| \le 2r) \ge \exp(-\phi(r) - v(x, r)^2/2). \tag{13.2}$$

Indeed, for any h such that $\|x - h\| \le r$ the inclusion

$$\{y : \|y - x\| \le 2r\} \supset \{y : \|y - h\| \le r\}$$

holds. Borell inequality for shifted sets (5.3) yields

$$\mathbb{P}\left(\|Y - x\| \le 2r\right) \ge \mathbb{P}\left(\|Y - h\| \le r\right)$$
$$\ge \mathbb{P}\left(\|Y\| \le r\right)\exp(-|h|^2_{H_P}/2)$$
$$= \exp(-\phi(r) - |h|^2_{H_P}/2).$$

By minimizing over h, we arrive at (13.2).

We will prove now that for any $\delta > 0$ the inequality

$$\mathbb{P}\left(v(X, r) \ge 2\sqrt{(2+\delta)\phi(r)}\right) \le \exp(-\phi(r)) \tag{13.3}$$

holds whenever r is small enough. Let $D = \{h : |h|_{H_P} \le 1\}$ be the dispersion ellipsoid of measure P, and let $U = \{x : \|x\| \le 1\}$ denote the unit ball of space $\mathscr{X}$. Then isoperimetric inequality implies that for any $u > 0$ we have

$$\mathbb{P}\left(v(X, r) \ge u\right) = \mathbb{P}\left((X + rU) \cap uD = \emptyset\right)$$
$$= \mathbb{P}\left(X \notin rU + uD\right)$$
$$\le \widehat{\Phi}\left(\Phi^{-1}\left(P(rU)\right) + u\right)$$
$$= \widehat{\Phi}\left(\Phi^{-1}\left(\exp(-\phi(r))\right) + u\right).$$

Since

$$\Phi^{-1}(p) \ge -\sqrt{(2+\delta)|\ln p|}$$

whenever p is small enough, we obtain

$$\mathbb{P}\left(v(X, r) \ge u\right) \le \widehat{\Phi}\left(-\sqrt{(2+\delta)\phi(r)} + u\right).$$

By letting $u = 2\sqrt{(2+\delta)\phi(r)}$, we have

$$\mathbb{P}\left(v(X, r) \ge 2\sqrt{(2+\delta)\phi(r)}\right) \le \widehat{\Phi}\left(\sqrt{(2+\delta)\phi(r)}\right)$$
$$\le \exp\left(-(2+\delta)\phi(r)/2\right)$$
$$\le \exp(-\phi(r)),$$

and we arrive at (13.3).

Let us now pass to evaluation of probabilities. By (13.2), for any $\lambda, r > 0$, $x \in \mathscr{X}$ we have a bound

$$\mathbb{P}\left(\min_{j \le e^\lambda} \|Y_j - x\| \ge 2r\right) = \mathbb{P}\left(\|Y - x\| \ge 2r\right)^{[e^\lambda]}$$
$$= (1 - \mathbb{P}\left(\|Y - x\| \le 2r\right))^{[e^\lambda]}$$
$$\le \exp\left\{-\mathbb{P}\left(\|Y - x\| \le 2r\right)[e^\lambda]\right\}$$
$$\le \exp\left\{-\exp(-\phi(r) - v(x, r)^2/2)[e^\lambda]\right\}.$$

For the probabilities of the balls with random centers we have the following. By using (13.3) we obtain

$$\mathbb{P}\left(M_\lambda \geq 2r\right) \leq \exp\left\{-\exp(-\phi(r) - (2\sqrt{(2+\delta)\phi(r)})^2/2)[e^\lambda]\right\} + \exp(-\phi(r))$$
$$= \exp\left\{-\exp(-\phi(r) - 2(2+\delta)\phi(r))[e^\lambda]\right\} + \exp(-\phi(r)).$$

Plugging $\delta = 1/3$ and $r = \phi^{-1}(\lambda/6)$ in this inequality yields

$$\mathbb{P}\left(M_\lambda \geq 2\phi^{-1}(\lambda/6)\right) \to 0, \quad \text{as} \quad \lambda \to \infty.$$

$\square$

The relation between quantization error and probabilities of *randomly centered* small balls stated in [48, 50] turns out to be even more tight. The latter notion is specified as follows. Let X be a Gaussian vector with distribution P. Define the small deviation function with random centers by

$$\Psi(\omega, r) = -\ln P\{x : ||x - X(\omega)|| \leq r\}.$$

In other words, we choose a random center of a ball according to distribution P, then measure deviations with respect to this center. Anderson inequality yields

$$P\{x : ||x - X(\omega)|| \leq r\} \leq P\{x : ||x|| \leq r\}.$$

Therefore,

$$\Psi(\omega, r) \geq \phi(r).$$

On the other hand, one can show that for a.s. ω it is true that

$$\lim_{r \to 0} \frac{\Psi(\omega, r)}{2\phi(r/2)} \leq 1.$$

One can also show that under natural regularity condition $\phi(r/2) \leq C\phi(r)$ small deviation function with random centers has a deterministic equivalent, i.e. there exists a non-increasing function $\phi_*(r)$ such that

$$\lim_{r \to 0} \frac{\Psi(\omega, r)}{\phi_*(r)} = 1 \quad \text{in probability.}$$

It is clear that the functions $\phi(r)$ and $\phi_*(r)$ have the same growth rate, as $r \searrow 0$. However, the exact relation between them is unknown. Even for Wiener process in space $\mathbb{C}[0, 1]$ we do not know the constant a in the formula $\phi_*(r) \sim a r^{-2}$. It is only known that a exists and $\frac{\pi^2}{4} \leq a \leq \pi^2$. The key role of the function $\phi_*(r)$ in quantization theory for Gaussian processes is demonstrated by the following theorem [50].

Theorem 13.3 *Assume that function ϕ_* satisfies assumption* (13.1). *Then for any* $p > 0$ *it is true that*

$$\lim_{\lambda \to \infty} \frac{D(\lambda, p)}{\phi_*^{-1}(\lambda)} = 1.$$

One can find many other interesting results and algorithms on quantization of Gaussian processes in the works of Graf, Luschgy and Pagès, see [78, 127–131] and references therein.

14 Invitation to Further Reading

These short lecture notes by no means aimed to provide a complete account of immense research field in pure and applied mathematics related to Gaussian processes.

Among the theoretical subjects for further reading we must recommend the majorizing measures and generic chaining, a powerful tool elaborated by Fernique and Talagrand for describing the boundedness and continuity of Gaussian processes. This technique is very impressive in its ability to solve difficult and critical cases. Being rather implicit, it was considered as difficult and obscure for some time but more accessible presentations were eventually found. There is an extensive first-hand literature [67, 68, 168, 173, 174]. See also [117] and especially in Chap. 6 of [105].

A large and important literature is devoted to the study of the *functionals* of Gaussian processes, including differential (Malliavin) calculus [22, 23, 43], and polynomial expansions (Wiener chaos) [105, 107, 144].

Concerning particular geometric features of multi-parameter and multi-dimensional random fields, the monographs by Adler and Taylor [2] and Khosh-nevisan [91] are specially recommended.

One can taste a variety of recent applications from the books and surveys such as Mandjes [135] and Willinger et al. [183] (models of communication networks) Rasmussen and Williams [150] (Machine Learning), van der Vaart and van Zanten [177] (prior models in Bayesian Statistics).

An interested reader will find further information on the theory of Gaussian processes in the monographs by Bogachev [22], Fernique [68], Kuo [99], Lifshits [117], Rozanov [151], and lectures by Adler [1] and Ledoux [105]. In more special aspects, the close themes are considered in the books of Ibragimov and Rozanov [84], Janson [87], Hida and Hitsuda [82].

References

1. Adler, R.J.: An introduction to continuity, extrema and related topics for general Gaussian processes. IMS Lecture Notes, Institute of Mathematical Statistics, vol. 12, Hayword (1990)
2. Adler, R.J., Tailor, J.E.: Random Fields and Their Geometry. Springer, New York (2007)

3. Anderson, T.W.: The integral of symmetric unimodal function. Proc. Amer. Math. Soc. **6**, 170–176 (1955)
4. Anderson, T.W., Darling, D.A.: Asymptotic theory of certain "goodness of fit" criteria based on stochastic processes. Ann. Math. Stat. **23**, 193–212 (1952)
5. Aronszajn, N.: Theory of reproducing kernels. Trans. Amer. Math. Soc. **68**, 337–404 (1950)
6. Artstein, S., Milman, V.D., Szarek, S.J.: Duality of metric entropy. Ann. Math. **159**, 1313–1328 (2004)
7. Aurzada, F.: Lower tail probabilities of some random sequences in l_p. J. Theor. Probab. **20**, 843–858 (2007)
8. Aurzada, F., Ibragimov, I.A., Lifshits, M.A., van Zanten, J.H.: Small deviations of smooth stationary Gaussian processes. Theor. Probab. Appl. **53**, 697–707 (2008)
9. Aurzada, F., Lifshits, M.A.: Small deviations of Gaussian processes via chaining. Stoch. Proc. Appl. **118**, 2344–2368 (2008)
10. Baldi, P., Ben Arous, G., Kerkacharian, G.: Large deviations and Strassen theorem in Hölder norm. Stoch. Proc. Appl. **42**, 171–180 (1992)
11. Bardina, X., Es-Sebaiy, K.: An extension of bifractional Brownian motion. Commun. Stochast. Anal. **5**, 333–340 (2011)
12. Barthe, F.: The Brunn–Minkowskii theorem and related geometric and functional inequalities. In: Proceedings of the International Congress of Mathematicians, vol. II, pp. 1529–1546. EMS, Zürich (2006)
13. Barthe, F., Huet, N.: On Gaussian Brunn–Minkowskii inequalities. Stud. Math. **191**, 283–304 (2009)
14. Bass, R.F., Pyke, R.: Functional law of the iterated logarithm and uniform central limit theorem for processes indexed by sets. Ann. Probab. **12**, 13–34 (1984)
15. Belinsky, E.S.: Estimates of entropy numbers and Gaussian measures for classes of functions with bounded mixed derivative. J. Approx. Theory **93**, 114–127 (1998)
16. Beghin, L., Nikitin, Ya.Yu, Orsinger, E.: Exact small ball constants for some Gaussian processes under L_2-norm. J. Math. Sci. **128**, 2493–2502 (2005)
17. Belyaev, Yu.K: Local properties of sample functions of stationary Gaussian processes. Theor. Probab. Appl. **5**, 117–120 (1960)
18. Bilyk, D., Lacey, M., Vagarshakyan, A.: On the small ball inequality in all dimensions. J. Funct. Anal. **254**, 2470–2502 (2008)
19. Bingham, N.H.: Variants on the law of the iterated logarithm. Bull. Lond. Math. Soc. **18**, 433–467 (1986)
20. Bobkov, S.G.: Extremal properties of half-spaces for log-concave distributions. Ann. Probab. **24**, 35–48 (1996)
21. Bobkov, S.G.: A functional form of isoperimetric inequality for the Gaussian measure. J. Funct. Anal. **135**, 39–49 (1996)
22. Bogachev, V.I.: Gaussian measures. Ser. Mathematical Surveys and Monographs, vol. 62. AMS, Providence (1998)
23. Bogachev, V.I.: Differentiable Measures and the Malliavin Calculus. Ser. Mathematical Surveys and Monographs, vol. 164, AMS, Providence (2010)
24. Borell, C.: The Brunn–Minkowsi inequality in Gauss space. Invent. Math. **30**, 207–216 (1975)
25. Borell, C.: Convex measures on locally convex spaces. Ark. Mat. **12**, 239–252 (1974)
26. Borell, C.: Gaussian Radon measures on locally convex spaces. Math. Scand. **38**, 265–284 (1976)
27. Borell, C.: A note on Gaussian measures which agree on small balls. Ann. Inst. H. Poincaré Ser. B **13**, 231–238 (1977)
28. Borell, C.: The Ehrhard inequality. C. R. Math. Acad. Sci. Paris. **337**, 663–666 (2003)
29. Borell, C.: Inequalities of the Brunn–Minkowski type for Gaussian measure. Probab. Theory Rel. Fields. **140**, 195–205 (2008)
30. Borovkov, A.A., Mogulskii, A.A.: On probabilities of small deviations for random processes. Siberian Adv. Math. **1**, 39–63 (1991)

31. Borovkov, A.A., Ruzankin, P.S.: On small deviations of series of weighted random variables. J. Theor. Probab. **21**, 628–649 (2008)
32. Bulinski, A.V., Shiryaev, A.N.: Theory of Random Processes. Fizmatlit, Moscow (in Russian) (2003)
33. Burago, Yu.D., Zalgaller, V.A.: Geometric Inequalities. Springer, Berlin (1988)
34. Cameron, R.H., Martin, W.T.: Transformations of Wiener integrals under translations. Ann. Math. **45**, 386–396 (1944)
35. Cameron, R.H., Martin, W.T.: The Wiener measure of Hilbert neighborhoods in the space of real continuous functions. J. Math. Phys. **23**, 195–209 (1944)
36. Chung, K.L.: On maximum of partial sums of sequences of independent random variables. Trans. Amer. Math. Soc. **64**, 205–233 (1948)
37. Chentsov, N.N.: Lévy Brownian motion for several parameters and generalized white noise. Theor. Probab. Appl. **2**, 265–266 (1961)
38. Chernoff, H.: A measure of asymptotic efficiency for tests of hypothesis based on sums of observations. Ann. Math. Stat. **23**, 493–507 (1952)
39. Ciesielski, Z., Kerkyacharian, G., Roynette, B.: Quelques espaces fonctionnels associés á des processus gaussiens. Stud. Math. **107**, 173–203 (1993)
40. Cordero-Erausquin, D., Fradelizi, M., Maurey, B.: The *B*-conjecture for the Gaussian measure of dilates of symmetric convex sets and related problems. J. Funct. Anal. **44**, 410–427 (2004)
41. Cramér, H.: Sur un nouveau théorème-limite de la théorie de probabilités. Act. Sci. Indust. **736**, 5–23 (1938)
42. Csáki, E.: A relation between Chung's and Strassen's law of the iterated logarithm. Z. Wahrsch. verw. Geb. **54**, 287–301 (1980)
43. Davydov, Yu.A., Lifshits, M.A., Smorodina, N.V.: Local Properties of Distributions of Stochastic Functionals. Translations of Mathematical Monographs, vol. 173. AMS, Providence (1998)
44. bibunstructuredde Acosta, A.: Small deviations in the functional central limit theorem with applications to functional laws of the iterated logarithm. Ann. Probab. **11**, 78–101 (1983)
45. Deheuvels, P., Lifshits, M.A.: Strassen-type functional laws for strong topologies. Probab. Theory Rel. Fields **97**, 151–167 (1993)
46. Deheuvels, P., Lifshits, M.A.: Necessary and sufficient condition for the Strassen law of the iterated logarithm in non-uniform topologies. Ann. Probab. **22**, 1838–1856 (1994)
47. Dembo, A., Zeitouni, O.: Large Deviation Techniques and Applications, 2nd edn. Springer, New York (1998)
48. Dereich, S.: Small ball probabilities around random centers of Gaussian measures and applications to quantization. J. Theor. Probab. **16**, 427–449 (2003)
49. Dereich, S., Fehringer, F., Matoussi, A., Scheutzow, M.: On the link between small ball probabilities and the quantization problem for Gaussian measures on Banach spaces. J. Theor. Probab. **16**, 249–265 (2003)
50. Dereich, S., Lifshits, M.A.: Probabilities of randomly centered small balls and quantization in Banach spaces. Ann. Probab. **33**, 1397–1421 (2005)
51. Dereich, S., Scheutzow, M.: High-resolution quantization and entropy coding for fractional Brownian motion. Electron. J. Probab. **11** (28), 700–722 (2006)
52. Dudley, R.M.: The sizes of compact subsets of Hilbert space and continuity of Gaussian processes. J. Funct. Anal. **1**, 290–330 (1967)
53. Dudley, R.M.: Sample functions of Gaussian processes. Ann. Probab. **1**, 66–103 (1973)
54. Dudley, R.M., Feldman, J., Le Cam, L.: On seminorms, and probabilities, and abstract Wiener spaces. Ann. Math. **93**, 390–408 (1971)
55. Dunker, T., Kühn, T., Lifshits, M.A., Linde, W.: Metric entropy of integration operators and small ball probabilities for the Brownian sheet. J. Approx. Theory **101**, 63–77 (1999)
56. Dunker, T., Lifshits, M.A., Linde, W.: Small deviations of sums of independent variables. In: High Dimensional Probability, Progress in Probability, Birkhäuser, vol. 43, pp. 59–74. Basel (1998)

57. Dzhaparidze, K., van Zanten, J.H.: Krein's spectral theory and the Paley–Wiener expansion for fractional Brownian motion. Ann. Probab. **33**, 620–644 (2005)

58. Dzhaparidze, K., van Zanten, J.H., Zareba, P.: Representations of isotropic Gaussian random fields with homogeneous increments. J. Appl. Math. Stoch. Anal. **7731** (2006)

59. Ehrhard, A.: Symetrisation dans l'espace de Gauss. Math. Scand. **53**, 281–301 (1983)

60. Fatalov, V.R.: Constants in the asymptotics of small deviation probabilities for Gaussian processes and fields. Russ. Math. Surv. **58**, 725–772 (2003)

61. Fatalov, V.R.: Occupation times and exact asymptotics of small deviations of Bessel processes for L_p-norms with $p > 0$. Izv. Math. **71**, 721–752 (2007)

62. Fatalov, V.R.: On exact asymptotics for small deviations of a nonstationary Ornstein–Uhlenbeck process in the L_p-norm, $p \geq 2$. Mosc. Univ. Math. Bull. **62**, 125–130 (2007)

63. Fatalov, V.R.: Exact asymptotics of small deviations for a stationary Ornstein-Uhlenbeck process and some Gaussian diffusion processes in the L_p-norm, $2 \leq p \leq \infty$. Probl. Inf. Transm. **44**, 138–155 (2008)

64. Fatalov, V.R.: Small deviations for two classes of Gaussian stationary processes and L_p-functionals, $0 < p \leq \infty$. Probl. Inf. Transm. **46**, 62–85 (2010)

65. Fehringer, F. Kodierung von Gaußmassen. Ph.D. Thesis, Technische Universität, Berlin (2001)

66. Feller, W.: The general form of the so-called law of the iterated logarithm. Trans. Amer. Math. Soc. **54**, 373–402 (1943)

67. Fernique X.: Régularité des trajectoires des fonctions aléatoires gaussiennes. In: École d'Été de Probabilités de Saint-Flour, IV-1974. Lecture Notes in Mathematics, vol. 480, pp. 1–96. Springer, Berlin (1975)

68. Fernique, X.: Fonctions aléatoires gaussiennes, Vecteurs aléatoirs gaussiens. CRM, Montreal (1997)

69. Fill, J.A., Torcaso, F.: Asymptotic analysis via Mellin transforms for small deviations in L_2-norm of integrated Brownian sheets. Probab. Theory Rel. Fields **130**, 259–288 (2003)

70. Freidlin, M.I.: The action functional for a class of stochastic processes. Theor. Probab. Appl. **17**, 511–515 (1972)

71. Gao, F., Hannig, J., Lee, T.-Y., Torcaso, F.: Laplace transforms via Hadamard factorization with applications to small ball probabilities. Electron. J. Probab. **8**(13), 1–20 (2003)

72. Gao, F., Hannig, J., Torcaso, F.: Integrated Brownian motions and exact l_2-small balls . Ann. Probab. **31**, 1320–1337 (2003)

73. Gao, F., Hannig, J., Lee, T.-Y., Torcaso, F.: Exact L^2-small balls of Gaussian processes. J. Theor. Probab. **17**, 503–520 (2004)

74. Gao, F., Li, W.-V.: Small ball probabilities for the Slepian Gaussian fields. Trans. Amer. Math. Soc. **359**, 1339–1350 (2005)

75. Gardner, R.J.: The Brunn–Minkowskii inequality. Bull. Amer. Math. Soc. **39**, 355–405 (2002)

76. Gardner, R.J., Zvavitch, A.: Gaussian Brunn–Minkowskii inequalities. Trans. Amer. Math. Soc. **362**, 5333–5353 (2010)

77. Golosov, J.N., Molchan, G.M.: Gaussian stationary processes with asymptotic power spectrum. Dokl. Math. **10**, 134–139 (1969)

78. Graf, S., Luschgy, H.: Foundations of Quantization for Probability Distributions. Lecture Notes in Mathematics, vol. 1730. Springer, Berlin (2000)

79. Grill, K.: Exact rate of convergence in Strassen's law of iterated logarithm. J. Theor. Probab. **5**, 197–205 (1992)

80. Hargé, G.: A particular case of correlation inequality for the Gaussian measure. Ann. Probab. **27**, 1939–1951 (1999)

81. Hartman, P., Wintner, A.: On the law of the iterated logarithm. Amer. J. Math. **63**, 169–176 (1941)

82. Hida, T., Hitsuda, M.: Gaussian Processes. AMS, Providence (1993)

83. Houdré, C., Villa, J.: An example of infinite dimensional quasi-helix. Contemporary Mathematics, vol. 336, pp. 195–201. AMS (2003)

84. Ibragimov, I.A., Rozanov, Yu. A.: Gaussian Random Processes. Springer, Berlin (1978)

85. Jain, N.C., Pruitt, W.E.: Maximum of partial sums of independent random variables. Z. Wahrsch. verw. Geb. **27**, 141–151 (1973)
86. Jain, N.C., Pruitt, W.E.: The other law of the iterated logarithm. Ann. Probab. **3**, 1046–1049 (1975)
87. Janson, S.: Gaussian Hilbert Spaces. Cambridge University Press, Cambridge (1997)
88. Karol', A., Nazarov, A., Nikitin, Ya.: Small ball probabilities for Gaussian random fields and tenzor products of compact operators. Trans. Amer. Math. Soc. **360**, 1443–1474 (2008)
89. Khatri, C.G.: On certain inequalities for normal distributions and their applications to simultaneous confidence bounds. Ann. Math. Stat. **38**, 1853–1867 (1967)
90. Khintchine, A.: Über einen Satz der Wahrscheinlichkeitsrechnung. Fund. Math. **6**, 9–12 (1924)
91. Khoshnevisan, D.: Multiparameter Processes: An Introduction to Random Fields. Springer, New York (2002)
92. Kolmogorov, A.N.: Sulla determinazione empirica di una legge di distribuzione . Attuari G. Inst. Ital. **4**, 83–91 (1933)
93. Kolmogorov A.N., Tikhomirov. V.M.: ϵ-entropy and ϵ-capasity of sets in functional space. Amer. Math. Soc. Transl. 2(17):277–364 (1961) Translated from Uspehi Mat. Nauk 14 (1959) (2):3–86 (Russian)
94. Khinchin, A.Ya.: Asymptotic Laws of Probability Theory. ONTI, Moscow–Leningrad (1936)
95. Komlós, J., Major, P., Tusnády, G.: An approximation of partial sums of independent RV'-s and the sample DF.I. Z. Wahrsch. verw. Geb. **32**, 111–131 (1975)
96. Komlós, J., Major, P., Tusnády, G.: An approximation of partial sums of independent RV'-s and the sample DF.II. Z. Wahrsch. verw. Geb. **34**, 34–58 (1976)
97. Kuelbs J., Li W.V.: Some shift inequalities for Gaussian measures. In: Eberlein, E. et al. (eds.) High Dimensional Probability. Proceedings of the conference, Oberwolfach, Germany, August 1996. Progress in Probability, vol. 43, pp. 233–243. Birkhäuser, Basel (1998)
98. Kuelbs, J., Li, W.V.: Metric entropy and the small ball problem for Gaussian measures. J. Funct. Anal. **116**, 133–157 (1993)
99. Kuo, H.-H.: Gaussian Measures in Banach Spaces. Lecture Notes in Mathematics, vol. 418. Springer (1975)
100. Kühn, T., Linde, W.: Optimal series representation of fractional Brownian sheet. Bernoulli **8**, 669–696 (2002)
101. Kwapień, S., Sawa, J.: On some conjecture concerning Gaussian measures of dilatations of convex symmetric sets. Stud. Math. **105**, 173–187 (1993)
102. Latała, R.: A note on Ehrhard inequality. Stud. Math. **118**, 169–174 (1996)
103. Latała, R.: On some inequalities for Gaussian measures. In: Proceedings of the International Congress of Mathematicians, vol. II, pp. 813–821. Higher Education Press, Beijing (2002)
104. Latała, R., Oleszkiewitcz, K.: Gaussian measures of dilatations of convex symmetric sets. Ann. Probab. **27**, 1922–1938 (1999)
105. Ledoux, M.: Isoperimetry and Gaussian Analysis. Lecture Notes in Mathematics, vol. 1648, pp. 165–294. Springer (1996)
106. Ledoux, M.: Concentration of Measure Phenomenon. Mathematical Surveys and Monographs, vol. 89. AMS (2001)
107. Ledoux, M., Talagrand, M.: Probability in Banach Spaces. Springer, New York (1991)
108. Lei, P., Nualart, D.: A decomposition of the bifractional Brownian motion and some applications. Statist. Probab. Lett. **79**, 619–624 (2009)
109. Lévy, P.: Problemes concrets d'analyse fonctionnelle. Gautier-Villars, Paris (1951)
110. Lévy, P.: Processus stochastiques et mouvement brownien, 2nd edn. Gautier-Villars, Paris (1965)
111. Li, W.V.: A Gaussian correlation inequality and its applications to small ball probabilities. Electron. Commun. Probab. **4**, 111–118 (1999)
112. Li, W.V.: Small deviations for Gaussian Markov processes under the sup-norm. J. Theor. Probab. **12**, 971–984 (1999)

113. Li, W.V.: Small ball probabilities for Gaussian Markov processes under the L_p-norm. Stoch. Proc. Appl. **92**, 87–102 (2001)
114. Li, W.V., Linde, W.: Existence of small ball constants for fractional Brownian motions, C. R. Math. Acad. Sci. Paris **326**, 1329–1334 (1998)
115. Li, W.V., Linde, W.: Approximation, metric entropy and small ball estimates for Gaussian measures. Ann. Probab. **27**, 1556–1578 (1999)
116. Li, W.V., Shao, Q.-M.: Gaussian processes: inequalities, small ball probabilities and applications. In: Rao, C.R., Shanbhag, D. (eds.) Stochastic Processes: Theory and Methods, Handbook of Statistics, vol. 19, pp. 533–597. North-Holland, Amsterdam (2001)
117. Lifshits, M.A.: Gaussian Random Functions. Kluwer, Dordrecht (1995)
118. Lifshits, M.A.: On representation of Lévy's fields by indicators. Theor. Probab. Appl. **29**, 629–633 (1979)
119. Lifshits, M.A.: On the lower tail probabilities of some random series. Ann. Probab. **25**, 424–442 (1997)
120. Lifshits, M.A.: Asymptotic behavior of small ball probabilities. In: Grigelionis, B. (ed.) Theory of Probability and Mathematical Statistics. Proceedings of VII International Vilnius Conference, 1998, pp. 453–468. VSP/TEV, Vilnius (1999)
121. Lifshits, M.A.: Bibliography on small deviation probabilities. http://www.proba.jussieu.fr/pageperso/smalldev/biblio.html
122. Lifshits, M.A., Linde, W.: Approximation and entropy numbers of Volterra operators with application to Brownian Motion. Mem. Amer. Math. Soc. **157**(745), 1–87 (2002)
123. Lifshits, M.A., Linde, W., Shi, Z.: Small deviations of Riemann–Liouville processes in L_q-norms with respect to fractal measures. Proc. Lond. Math. Soc. **92**, 224–250 (2006)
124. Lifshits, M.A., Linde, W., Shi, Z.: Small deviations of Gaussian random fields in L_q-spaces. Electron. J. Probab. **11**(46), 1204–1223 (2006)
125. Lifshits, M.A., Simon, T.: Small deviations for fractional stable processes. Ann. Inst. H. Poincaré, Ser. B **41**, 725–752 (2005)
126. Linde, W., Pietsch, A.: Mappings of Gaussian measures in Banach spaces. Theor. Probab. Appl. **19**, 445–460 (1974)
127. Luschgy, H., Pagès, G.: Functional quantization of Gaussian processes. J. Funct. Anal. **196**, 486–531 (2002)
128. Luschgy, H., Pagès, G.: Sharp asymptotics of the functional quantization problem for Gaussian processes. Ann. Probab. **32**, 1574–1599 (2004)
129. Luschgy, H., Pagès, G.: High-resolution product quantization for Gaussian processes under sup-norm distortion. Bernoulli **13**, 653–671 (2007)
130. Luschgy, H., Pagès, G.: Expansion for Gaussian processes and Parseval schemes. Electron. J. Probab. **14**, 1198–1221 (2009)
131. Luschgy, H., Pagès, G., Wilbertz, B.: Asymptotically optimal quantization schemes for Gaussian processes. ESAIM: Probab. Stat. **14**, 93–116 (2010)
132. Major, P.: An improvement of Strassen's invariance principle. Ann. Probab. **7**, 55–61 (1979)
133. Malyarenko, A.: An optimal series expansion of the multiparameter fractional Brownian motion. J. Theor. Probab. **21**, 459–475 (2008)
134. Mandelbrot, B.B., van Ness, J.: Fractional Brownian motions, fractional noises and applications. SIAM Rev. **10**, 422–437 (1968)
135. Mandjes, M.: Large Deviations for Gaussian Queues: Modelling Communication Networks. Wiley, Chichester (2007)
136. Marinucci, D., Robinson, P.M.: Alternative forms of fractional Brownian motion. J. Stat. Plan. Infer. **80**, 111–122 (1999)
137. Nazarov, A.I.: On the sharp constant in the small ball asymptotics of some Gaussian processes under L_2-norm. J. Math. Sci. **117**, 4185–4210 (2003)
138. Nazarov, A.I.: Exact L_2-small ball asymptotics of Gaussian processes and the spectrum of boundary-value problems. J. Theor. Probab. **22**, 640–665 (2009)

139. Nazarov, A.I.: Nikitin, Ya.Yu.: Exact L_2-small ball behavior of integrated Gaussian processes and spectral asymptotics of boundary value problems. Probab. Theory Rel. Fields **129**, 469–494 (2004)

140. Nazarov, A.I., Nikitin, Ya.Yu.: Logarithmic L_2-small ball asymptotics for some fractional Gaussian processes. Theor. Probab. Appl. **49**, 695–711 (2004)

141. Nazarov, A.I., Pusev, R.S.: Exact small deviation asymptotics in L_2-norm for some weighted Gaussian processes. J. Math. Sci. **163**, 409–429 (2009)

142. Paley, R.E.A.C., Wiener, N.: Fourier Transforms in the Complex Domain. In: American Mathematical Society Colloquium Publications, vol. 19. AMS, New York (1934). Reprint: AMS, Providence (1987)

143. Park, W.J.: On Strassen version of the law of the iterated logarithm for two-parameter Gaussian processes. J. Multivariate Anal. **4**, 479–485 (1974)

144. Peccati, G., Taqqu, M.S.: Wiener Chaos: Moments, Cumulants and Diagrams. Springer, Berlin (2011)

145. Petrov, V.V.: Limit Theorems of Probability Theory. Sequences of Independent Random Variables. Clarendon Press, Oxford (1995)

146. Petrovskii, I.G: Über das Irrfahrtproblem. Math. Ann. **109**, 425–444 (1934)

147. Pisier, G.: Conditions d'entropie assurant la continuité de certains processus et applications à l'analyse harmonique. Seminaires d'analyse fonctionnelle, Exp. 13–14 (1980).

148. Pisier, G.: The Volume of Convex Bodies and Banach Space Geometry. Cambridge University Press, Cambridge (1989)

149. Pitt, L.: A Gaussian correlation inequality for symmetric convex sets. Ann. Probab. **5**, 470–474 (1977)

150. Rasmussen, C.E., Williams, C.K.I: Gaussian Processes for Machine Learning. MIT Press, Cambridge (2006)

151. Rozanov, Yu. A.: Gaussian Infinite-Dimensional Distributions. Proceedings of Steklov Institute Mathematics, vol. 108. AMS, Providence (1971)

152. Sakhanenko, A.I.: Rate of convergence in the invariance principle for variables with exponential moments that are not identically distributed. In: Advances in Probability Theory: Limit Theorems for Sums of Random Variables, pp. 2–73. Optimization Software, New York (1985)

153. Sato, H.: Souslin support and Fourier expansion of a Gaussian Radon measure. In: A. Beck (ed.) Probability in Banach Spaces III. Proceedings of International Conference, Medford, USA, 1980. Lecture Notes in Mathematics, **860**, pp. 299–313. Springer, Berlin (1981)

154. Schechtman, G., Schlumprecht, T., Zinn, J.: On the Gaussian measure of intersection. Ann. Probab. **26**, 346–357 (1998)

155. Schilder, M.: Some asymptotic formulas for Wiener integrals. Trans. Amer. Math. Soc. **125**, 63–85 (1966)

156. Šidák, Z.: Rectangular confidence regions for the means of multivariate normal distributions. J. Amer. Statist. Assoc. **62**, 626–633 (1967)

157. Skorokhod, A.V.: Integration in Hilbert space. Ergebnisse der Mathematik und ihrer Grenzgebiete, vol. 79. Springer, Berlin (1974)

158. Steiner, J.: Einfacher Beweis der isoperimetrischen Hauptsätze. J. reine angew. Math. **18**, 281–296 (1838)

159. Stolz, W.: Une méthode élémentaire pour l'évaluation des petites boules browniennes. C. R. Math. Acad. Sci. Paris **316**, 1217–1220 (1993)

160. Stolz, W.: Small ball probabilities for Gaussian processes under non-uniform norms. J. Theor. Probab. **9**, 613–630 (1996)

161. Strassen, V.: An invariance principle for the law of the iterated logarithm. Z. Wahrsch. verv. Geb. **3**, 211–226 (1964)

162. Sudakov, V.N.: Gaussian measures, Cauchy measures and ε-entropy. Dokl. Math. **10**, 310–313 (1969)

163. Sudakov, V.N.: Gaussian random processes and measures of solid angles in a Hilbert space. Dokl. Math. **12**, 412–415 (1971)

164. Sudakov, V.N.: Geometric Problems in the Theory of Infinite-Dimensional Probability Distributions. Proc. Steklov Inst. Math. **141**, 1–178 (1976)
165. Sudakov, V.N., Tsirelson, B.S.: Extremal properties of half-spaces for spherically invariant measures. J. Sov. Math. **9**, 9–18 (1978) (Russian original: Zap. Nauchn. Semin. POMI, **41**, 14–41 (1974))
166. Talagrand, M.: Regularity of Gaussian processes. Acta Math. **159**, 99–149 (1987)
167. Talagrand, M.: On the rate of clustering in Strassen's law of the iterated logarithm for Brownian motion. In: Probability in Banach space. VIII, Progress in Probability, **30**, pp. 339–347. Birkhäuser, Basel (1992)
168. Talagrand, M.: Simple proof of the majorizing measure theorem. Geom. Funct. Anal. **2**, 118–125 (1992)
169. Talagrand, M.: New Gaussian estimates for enlarged balls. Geom. Funct. Anal. **3**, 502–526 (1993)
170. Talagrand, M.: The small ball problem for the Brownian sheet. Ann. Probab. **22**, 1331–1354 (1994)
171. Talagrand, M.: Concentration of measure and isoperimetric inequalities in product spaces. Publ. Math. de l'I.H.E.S. **81**, 73–205 (1995)
172. Talagrand, M.: A new look at independence. Ann. Probab. **24**, 1–34 (1996)
173. Talagrand, M.: Majorizing measures: the generic chaining. Ann. Probab. **24**, 1049–1103 (1996)
174. Talagrand, M.: The Generic Chaining: Upper and Lower Bounds for Stochastic Processes. Springer, Berlin (2005)
175. Temliakov, V.N.: Estimates of asymptotic characteristics of classes of functions with bounded mixed derivative and difference. Proc. Steklov Inst. Math. **12**, 161–197 (1990)
176. van der Vaart, A.W., van Zanten, J.H: Rates of contraction of posterior distributions based on Gaussian process priors. Ann. Statist. **36**, 1435–1463 (2008)
177. van der Vaart, A.W., van Zanten, J.H.: Bayesian inference with rescaled Gaussian process priors. Electron. J. Statist. **1**, 433–448 (2007)
178. Varadhan, S.R.S.: Asymptotic probabilities and differential equations. Comm. Pure Appl. Math. **19**, 261–286 (1966)
179. Weber, M.: Entropie métrique et convergence presque partout, vol. 58. Travaux en Cours. Hermann, Paris (1998)
180. Weber, M.: Dynamical Systems and Processes. IRMA Lectures in Mathematics and Theoretical Physics, vol. 14. EMS, Zürich (2009)
181. Wentzell, A.D.: Theorems on the action functional for Gaussian random functions. Theor. Probab. Appl. **17**, 515–517 (1972)
182. Wentzell, A.D.: A Course in the Theory of Stochastic Processes. McGraw–Hill, New York (1981)
183. Willinger, W., Paxson, V., Riedli, R.H., Taqqu, M.S.: Long range dependence and data network traffic. In: Theory and Applications of Long Range Dependence, pp. 373–408. Birkhäuser, Basel (2003)

Index

A
Absolutely continuous measure, 34
Admissible direction, 34
Admissible shift, 34
Anderson inequality, 47–48, 52, 91, 109

B
B-concavity, 49
Belyaev alternative, 77
Bifractional Brownian motion, 13, 17
Borell inequality for shifted sets, 38, 89
Brownian bridge, 11, 16, 31, 81, 101
Brownian sheet, 11, 18, 30, 69, 104
Brunn-Minkowski inequality, 46

C
Cameron–Martin formula, 35
Cameron–Martin space, 28
Cameron-Martin theorem, 34, 37, 53
Chaining, 74, 85, 110
Chentsov construction, 19
Chung's FLIL, 69
Concentration principle, 44, 61
Correlation conjecture, 50
Covariance operator, 3, 5
Covering number, 72, 83, 90, 93
Cramér-Chernoff theorem, 54

D
Deviation function, 55–57
Dictionary, 106
Discretization problem, 106

Dispersion ellipsoid, 24, 43, 63, 80, 89, 90, 92, 93
Duality conjecture, 92
Duality of metric entropy, 92
Dudley integral, 74
Dudley upper bound, 74

E
Ehrhard concave measure, 47
Enlargement, 39–41, 43, 53
Expectation, 5

F
Factorization theorem, 26–30, 105
Fernique criterion, 77
Fernique-Sudakov comparison theorem, 77
Fractional Brownian motion, 10, 16, 17, 29, 70, 79, 84, 92, 104
Functional law of the iterated logarithm (FLIL), 62, 63

G
GB-set, 80
GC-set, 80
Gaussian large deviation principle, 57, 67
Gaussian Markov process, 11
Gaussian random process, 8
Gaussian random variable, 1
Gaussian stationary process, 20, 77, 79, 95
Gaussian vector in Hilbert space, 7, 25, 93
Gaussian vector associated to an operator, 105

G (*cont.*)
Gaussian white noise, 13, 14
Gaussian chaining, 110

H
Haar base, 103
Haar-Schauder expansion, 103
Half-space, 41
Hartman-Wintner LIL, 61
High resolution quantization, 106

I
Integral representation, 15
Isonormal random function, 80
Isoperimetric function, 39
Isoperimetric inequality, 38, 40, 41, 43, 91, 108
Isoperimetric problem, 38

K
Karhunen–Loève expansion, 7, 93, 102
Khatri-Šidák inequality, 51, 86
Kernel, 24
Khinchin LIL, 61
Kiefer field, 12
KMT-construction, 70
Komlós-Major-Tusnády strong invariance
        principle, 70

L
Large deviation principle, 55–57
Large deviation rate, 55
Law of the iterated logarithm (LIL), 61
Lévy construction for Wiener process, 103
Lévy inequality, 99
Lévy's Brownian function, 12, 19
Limit set, 62
Lipschitz functional, 44, 45
ℓ-norm, 105
Logarithmically concave measure, 46

M
Machine learning, 110
Malliavin calculus, 110
Major strong invariance principle, 71
Majorizing measures, 110
Median, 44
Measurable linear functional, 23

Metric capacity, 73, 89
Metric entropy, 72, 90
Multivariate Wiener process, 69

N
Natural distance, 73
Normal distribution in $\mathbb{R}$, 1
Normal random variable, 1

O
Ornstein–Uhlenbeck process, 11, 22, 30

P
Packing number, 72
Pisier theorem, 75
Poincaré lemma, 40
Polar representation of a kernel, 33

Q
Quantization problem, 106

R
Random process, 8
Random walk, 71
Randomly centered small balls, 109
Range, 84
Regular set, 56
Reproducing kernel Hilbert space (RKHS), 32
Riemann–Liouville process, 17, 92

S
Sakhanenko strong invariance principle, 71
Schauder functions, 103
Shift representation, 22
Self-similarity, 9, 10, 12, 13, 18, 62, 66, 69,
        70, 81
Shift of a measure, 34
Small ball problem, 80
Small deviation function, 88, 107
Small deviation function with random centers,
        109
Small deviation problem, 80
Small deviation rate, 81
Small deviation constant, 81
Spectral representation, 11, 15, 20, 21
S-property, 49

Stability, 1, 3, 4, 7, 52
Standard Gaussian random vector, 2
Standard Gaussian measure in $\mathbb{R}^\infty$, 6, 24
Standard normal distribution, 2
Stochastic approximation numbers, 105
Strassen ball, 63
Strassen FLIL, 63, 72
Strassen strong invariance principle, 71
Strong invariance principle, 70
Sudakov lower bound, 79

T
Talagrand bound, 84

U
Usual assumptions, 5, 6, 23, 105

W
Weak correlation inequality, 51
White noise integral, 14
Wiener chaos, 110
Wiener-Chentsov random field, 11, 104
Wiener process, 8, 9, 15, 27, 35, 61, 62, 81, 101–104